Structural Analysis I
Lecture Notes

Peter I. Kattan

Petra Books
www.PetraBooks.com

Peter I. Kattan, PhD

Correspondence about this book may be sent to the author at one of the following two email addresses:

pkattan@petrabooks.com

info@petrabooks.com

In Loving Memory of My Parents

Structural Analysis I Lecture Notes

Preface

These are the handwritten notes for the course "Structural Analysis I" that was taught at Applied Science University in the period 1996-1998. These notes are based on the book "Structural Analysis" by Alexander Chajes, Second Edition. This book is out of print at the present time. Sutdents find these notes useful and helpful in their studies and the author is glad to make them available to students worldwide. It is hoped that these notes will also revive the Chajes book.

The notes are freely available on the internet for free download as individual chapters. But their existence in one single volume is available in this book format only. The notes have no chapter 1. They start immediately with chapter 2 on calculation of reactions.

November 2021   Peter I. Kattan

Contents

Structural Analysis I

Chapter 2: Calculation of Reactions

Introduction:

* The first step consists of drawing a <u>free-body diagram</u>.

This is a simplified picture of the structure, isolated from its supports, on which are shown all the external forces that act on the structure. These forces include the <u>applied loads</u> and the <u>reactions</u> that the support exerts on the structure.

Example:

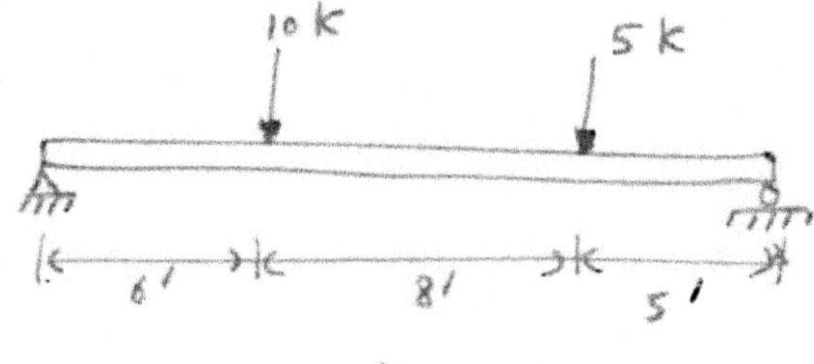

Free-body diagram

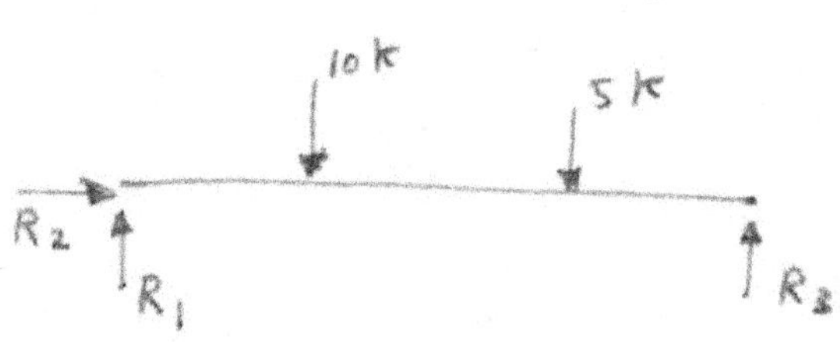

---

* <u>Equations of Equilibrium:</u>

In three dimensions: $\Sigma F_x = 0$, $\Sigma F_y = 0$, $\Sigma F_z = 0$

$$\Sigma M_x = 0, \quad \Sigma M_y = 0, \quad \Sigma M_z = 0$$

In two dimensions: $\boxed{\Sigma F_x = 0, \quad \Sigma F_y = 0, \quad \Sigma M_o = 0}$

---

* <u>Types of Supports and Restraints:</u>

(1) Roller:

(2) Hinged Support :

(3) Fixed Support :

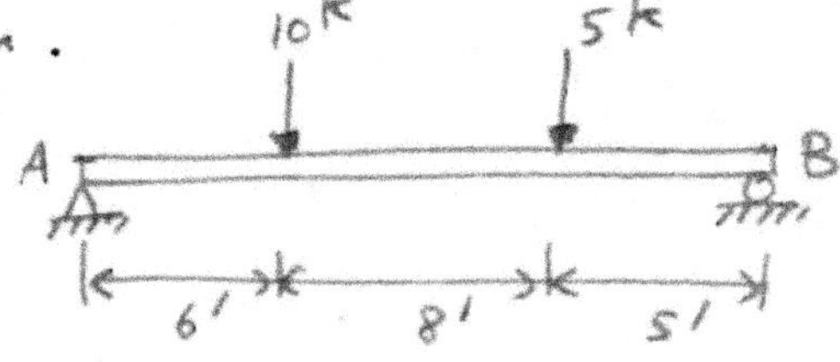

(4) Cable

<u>Example</u> :

     Find the reactions for the beam shown.

<u>Solution</u> :

(1) Draw the free-body diagram :

(2) Equations of equilibrium :

$$\Sigma F_x = 0 \implies R_2 = 0$$

$$\Sigma m_A = 0 \curvearrowright$$

$$\implies R_3 (19) - 5 (14) - 10 (6) = 0$$

$$\implies R_3 = 6.84 \text{ K}$$

$$\Sigma F_y = 0 \implies R_1 + R_3 - 10 - 5 = 0 \implies R_1 + 6.84 - 15 = 0$$

$$\implies R_1 = 8.16 \text{ k}$$

(3) Check the calculations :

$$\Sigma m_B = 0 \implies \curvearrowright \implies 5(5) + 10(13) - R_1(19) = 0$$

$$\implies 25 + 130 - 19(8.16) = 0$$

$$\implies 0 = 0 \checkmark$$

(4) Show all the reactions on a free-body diagram of the structure

<u>Stability and Determinacy</u> :

* A <u>determinate structure</u> is a structure for which all the unknown reactions can be determined using the equations of equilibrium.

<u>Example</u> :

   - Three unknown reactions

   - Three equations of equilibrium

∴ The structure is <u>determinate</u>.

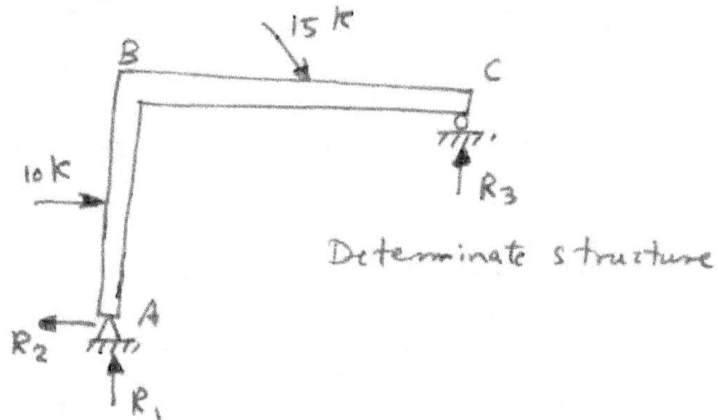

<u>Example</u> :

   - Four unknown reactions

   - Three equations of equilibrium

∴ The structure is <u>indeterminate</u>.

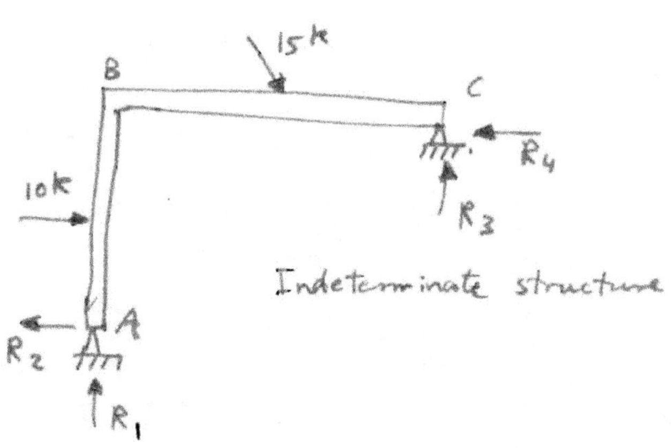

* An <u>indeterminate</u> structure is a structure which possesses more unknown reactions than equations of equilibrium.

<u>Example</u> :

   - There is nothing to prevent the structure from moving toward the right when subjected to the applied loads.

∴ The structure is <u>unstable</u>.

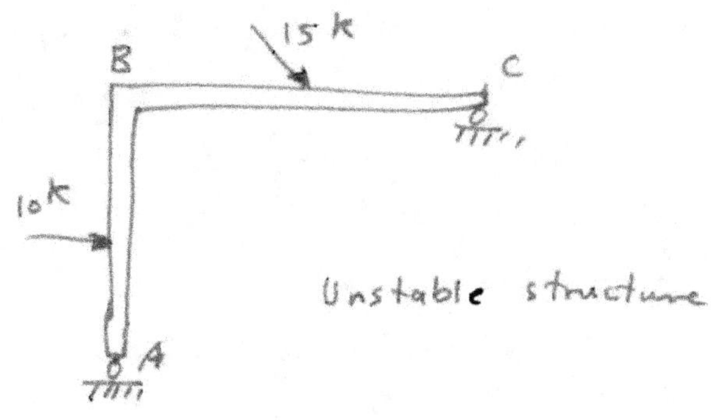

   * An <u>unstable</u> structure is a structure in which there are insufficient number of reactions to prevent motion from taking place.

* In general, a plane structure is both <u>stable</u> and <u>determinate</u> if it is supported by <u>three</u> reaction components that are neither parallel nor concurrent.

Example :
- The structure is stable as long as it is only subjected to a vertical load.
- the structure is unable to resist a horizontal load and is <u>unstable</u>.

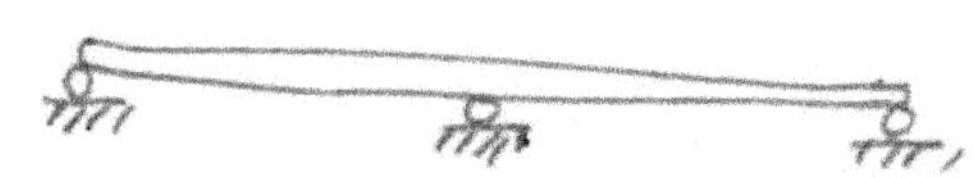

* Although at least <u>three</u> reaction components are necessary to provide stability, this number is <u>not</u> always sufficient.

Example :
- the structure has <u>three</u> reactions.
- the structure cannot resist a horizontal load because the reactions are all vertical.
- the structure is <u>unstable</u>.

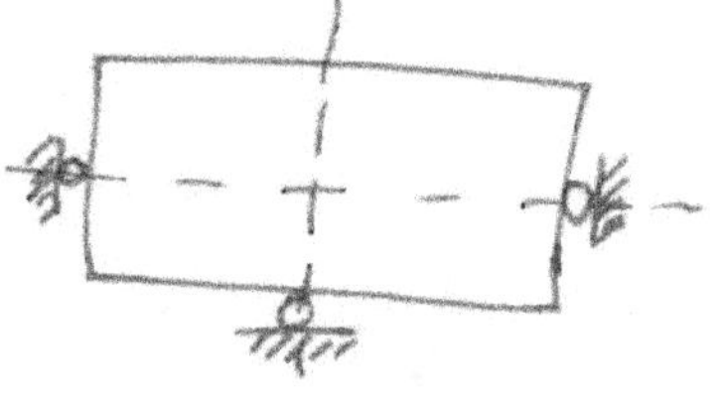

Example :
- The structure has <u>three</u> reactions.
- The lines of action of the three reactions all pass through a common point, making it impossible for them to resist an applied moment about this point.
- The structure is <u>unstable</u>.

Example :
- There are three unknown reactions.
- There are three equations of equilibrium.
- The structure is <u>determinate</u>.

Example :
- There are four unknown reactions.
- There are three equations of equilibrium.
- The structure is <u>indeterminate</u>.

Example :

- This is a __compound__ structure.

- There are four unknown reactions.

- There are four equations
  (3 equilibrium equations
   + 1 equation at the hinge)

- The structure is __determinate__.

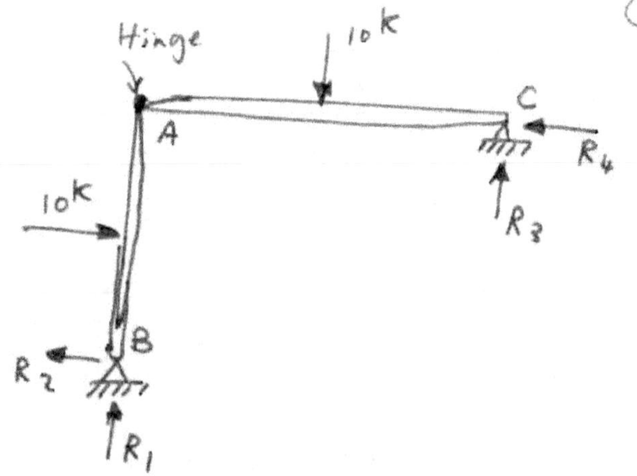

* __Structure with hinge at A__ :

- There are six unknowns

- There are six equations
  of equilibrium
   (three equations for
    each free-body diagram)

- The structure is __determinate__.

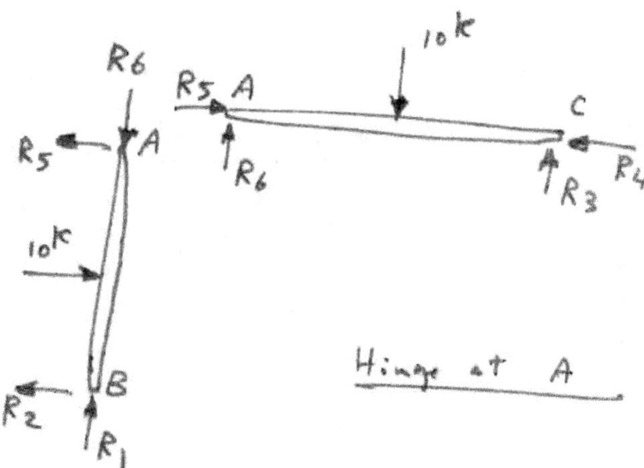

* __Structure without hinge at A__ :

- There are seven unknowns.

- There are six equations
  of equilibrium
   (three equations for each
    free-body diagram)

- The structure is __indeterminate__.

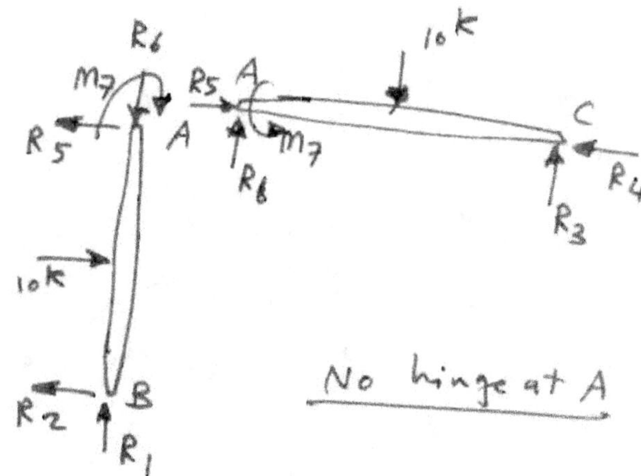

Calculation of Reactions : Simple Structures :

Example 2·1 :

Find the reactions for the structure shown.

Solution:

(1) Draw the free-body diagram

There are three unknowns.

There are three equations of equilibrium.

∴ the structure is <u>determinate</u>.

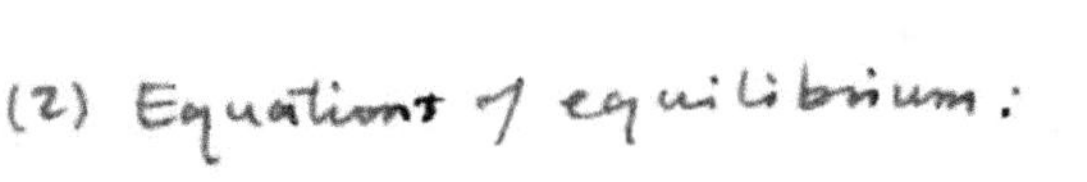

(2) Equations of equilibrium :

$$\Sigma F_x = 0 \implies R_1 + 20 - 36 = 0$$
$$\implies R_1 = 16^k$$

$$\Sigma m_A = 0 \;\; \circlearrowleft : \quad R_3(20) + 36(12) - 48(10) - 20(6) = 0$$
$$\implies R_3 = 8.4^k .$$

$$\Sigma F_y = 0 \implies R_2 + R_3 - 48 = 0 \implies R_2 + 8.4 - 48 = 0$$
$$\implies R_2 = 39.6^k .$$

(3) Check the calculations :

$$\Sigma M_B = 0 \;\; \circlearrowleft : \quad 48(10) + 20(6) + R_1(12) - R_2(20) = 0$$
$$48(10) + 20(6) + 16(12) - 39.6(20) = 0$$
$$0 = 0 \;\; \checkmark$$

(4) Show all the reactions on a free-body diagram of the structure:

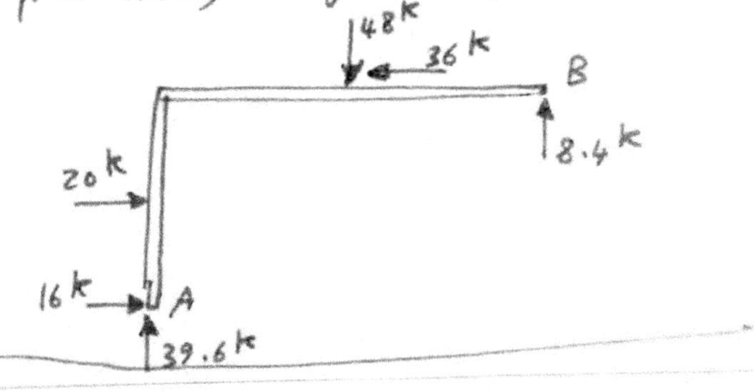

Example :

The reactions are <u>concurrent</u>.
(i.e. they intersect at the same point), point B.

∴ The structure is <u>unstable</u>.

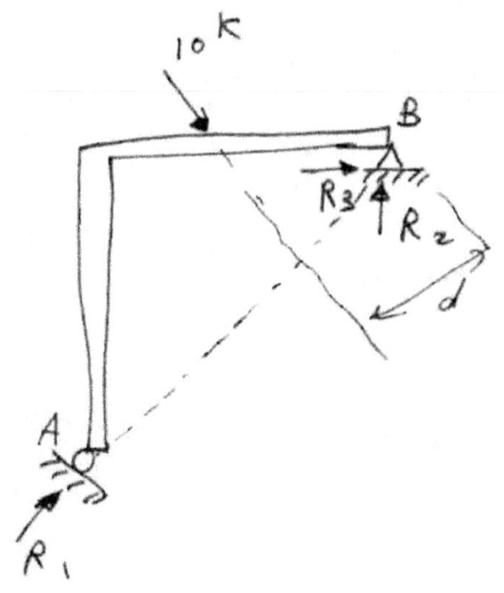

$+)\ \Sigma M_B = 0$: $10d \neq 0$

The equation of equilibrium is <u>not</u> satisfied.

the structure is n<u>ot</u> in equilibrium.

Example :

The reactions are <u>parallel</u>.

∴ The structure is <u>unstable</u>.

Also, it is <u>indeterminate</u>.

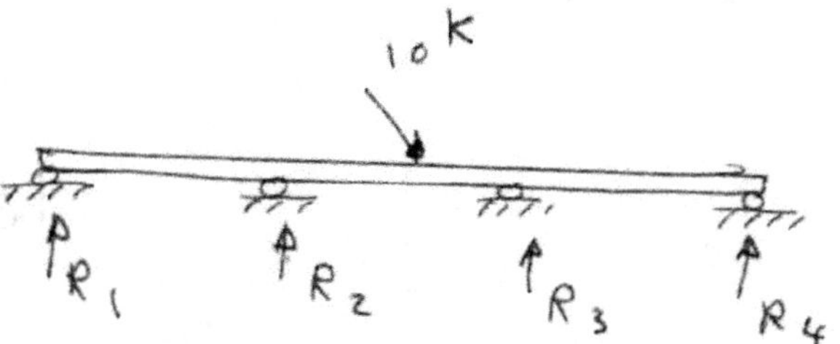

Calculation of Reactions: Compound Structures:

A **Compound Structure** is a structure made up of a number of members connected to one another in a non-rigid manner.

Example 2.2:

Find the reactions for the structure shown.

Solution:

(1) Draw the free-body diagram

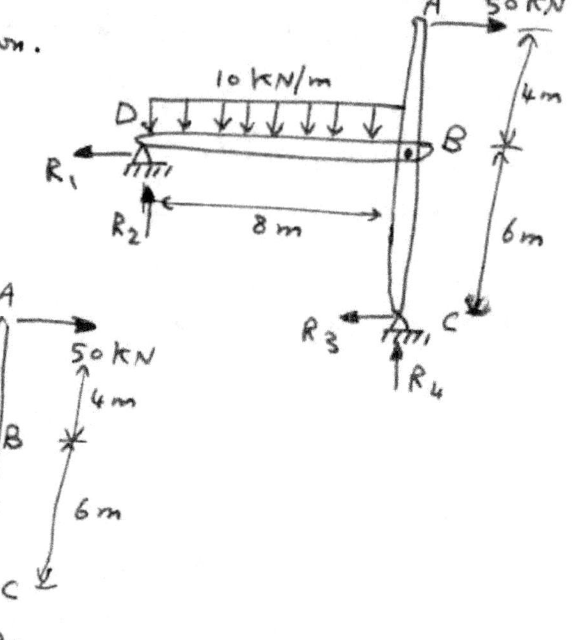

There are six unknowns. There are six equations of equilibrium.
∴ The structure is **determinate**.

(2) Equations of equilibrium:

For ABC↑ $\Sigma M_B = 0$ ↺ ⟹ $-50(4) + R_3(6) = 0$ ⟹ $R_3 = 33.3$ kN.

$\therefore R_3 = 33.3$ kN →

For the entire structure: $\Sigma M_D = 0$ ↺ :

$10(8)(4) + 50(4) + R_3(6) - R_4(8) = 0$

$10(8)(4) + 50(4) + (-33.3)(6) - 8R_4 = 0$ ⟹ $R_4 = 40$ kN ↑ .

$\Sigma F_x = 0$ ⟹ $-R_1 + 50 - R_3 = 0$ ⟹ $-R_1 + 50 - (-33.3) = 0$

⟹ $R_1 = 83.3$ kN ←

$\Sigma F_y = 0$ ⟹ $R_2 + R_4 - 10(8) = 0$ ⟹ $R_2 + 40 - 80 = 0$

⟹ $R_2 = 40$ kN ↑ .

(3) Check the calculations:

$\Sigma M_A = 0$ ↺ : $10(8)(4) - R_1(4) - R_2(8) - R_3(10) = 0$

$10(8)(4) - 83.3(4) - 40(8) - (-33.3)(10) = 0$ ⟹ $0 = 0$ ✓

(4) Show all the reactions on a free-body diagram of the structure: ⑧

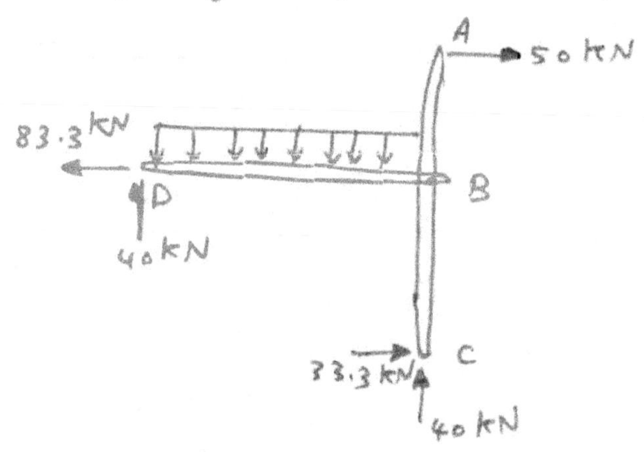

Example 2.3 :

Determine the reactions for the (cantilever structure) shown.

* The structure is
determinate:
(five unknowns,
five equations:
three equilibrium
equations + 2 equations
for the two hinges)

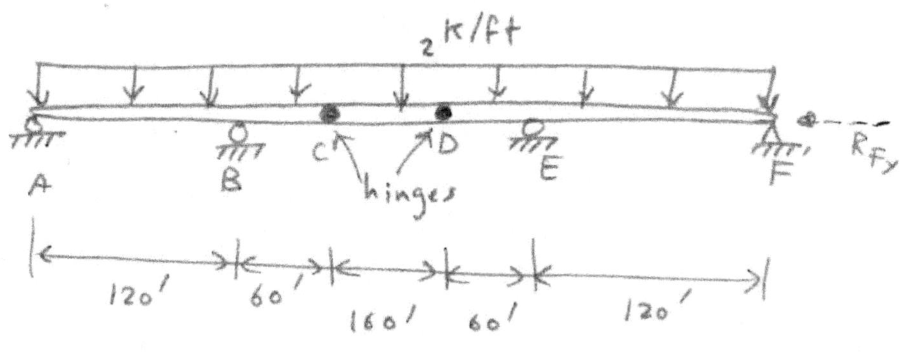

Solution :

(1) Draw the free-body diagram :

(2) Equations of equilibrium :

CD: $\Sigma F_y = 0 \Rightarrow R_C + R_D - 2(160) = 0$

     but $R_C = R_D$ from symmetry

     $\Rightarrow 2R_C - 2(160) = 0 \Rightarrow R_C = R_D = 160^k$.

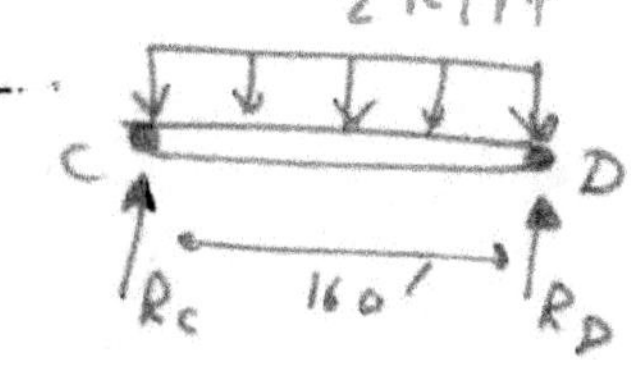

ABC:

     $\Sigma M_A = 0 \; \circlearrowright :$

     $-2(180)(90) - 160(180) + R_B(120) = 0$

     $\Rightarrow R_B = 510^k$.

     $\Sigma F_y = 0 \Rightarrow R_A + R_B - 160 - 2(180) = 0$

         $R_A + 510 - 160 - 2(180) = 0$

         $\Rightarrow R_A = 10^k$.

For the entire structure : $\Sigma F_x = 0 \Rightarrow R_{F_x} = 0$

∴ the structure is <u>symmetric</u> about its centerline :

∴ $R_F = R_A = 10^K \uparrow$

$R_E = R_B = 510^K \uparrow$

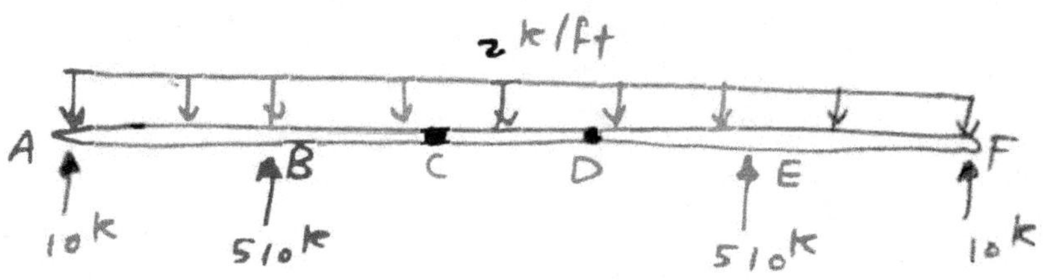

$\boxed{\text{Homework 1}}$: $\underline{2.1}$, $\underline{2.8}$, $\underline{2.12}$, $\underline{2.18}$, $\underline{2.24}$.

Chapter 3: Plane Trusses

3.1 Introduction:

* A __truss__ is a system of relatively slender members, arranged in the form of one or more triangles, which transfers loads by developing __axial forces__ in the members.

3.2 Simplifying Assumptions:

* The following assumption are used:

(1) The members are connected to each other at their ends by frictionless pins (joints); i.e. only a force and __no__ moment can be transferred from one member to another.

(2) External loads are applied to the truss __only__ at its __joints__.

(3) The centroidal axes of the members meeting at a joint all intersect at a common point; i.e. the point where the members are assumed to be pinned to one another.

__Conclusion__: the members of a truss are subjected to axial forces only.

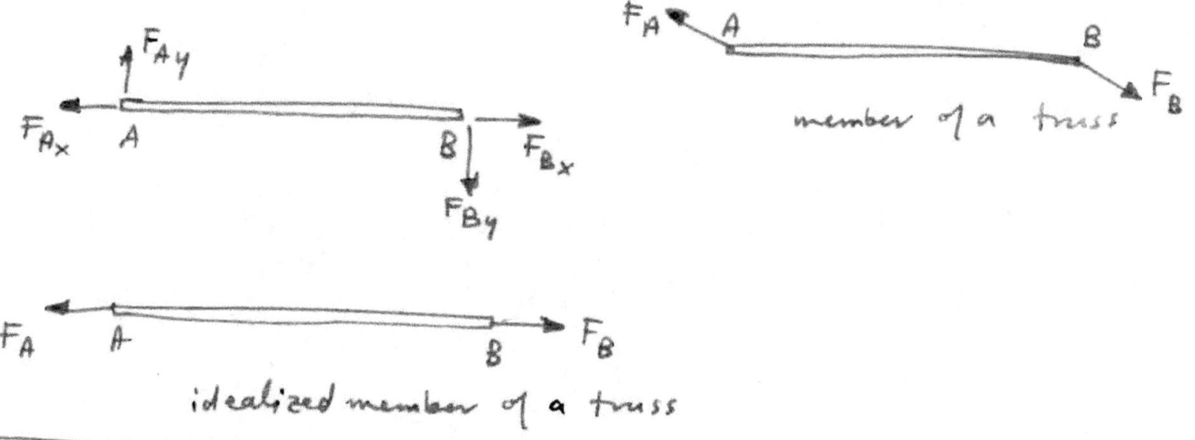

3.3 Basic Concepts:

* To analyze a truss ≡ to find the forces produced in its members by a set of external loads.

* There are two methods for analyzing a truss:
  (1) Method of Joints: we consider free bodies of individual joints.
  (2) Method of Sections: We pass a section that bisects the truss.

Example:

$$F_{AB,x} = F_{AB}\cos\alpha = F_{AB}\left(\frac{a}{c}\right)$$

$$F_{AB,y} = F_{AB}\sin\alpha = F_{AB}\left(\frac{b}{c}\right)$$

where $c = \sqrt{a^2+b^2}$. * There is no need to calculate the angle α.

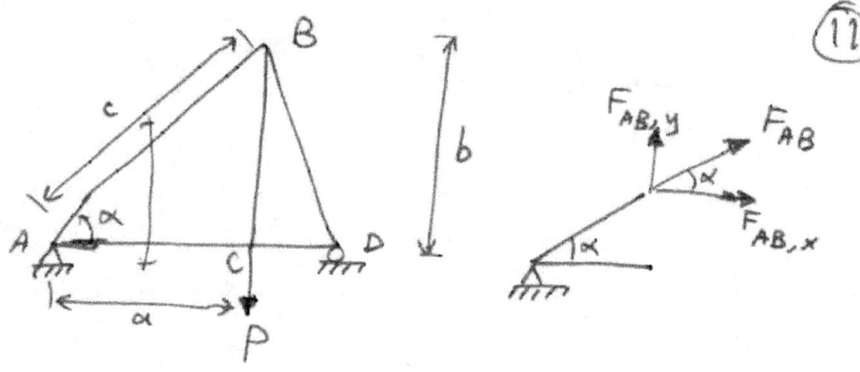

* Sign Convention:

The members forces can be either:

(1) tension (T), (+)

(2) compression (C), (−)

3.4 Method of Joints:

* A section is passed completely around a joint.
* Use the two equations of equilibrium: $\Sigma F_x = 0$, $\Sigma F_y = 0$.

Example:

$\Sigma F_x = 0$:

$\Rightarrow F_{AC} - F_{AB}\cos\alpha = 0$

$\Sigma F_y = 0$:

$\Rightarrow R_A - F_{AB}\sin\alpha = 0$

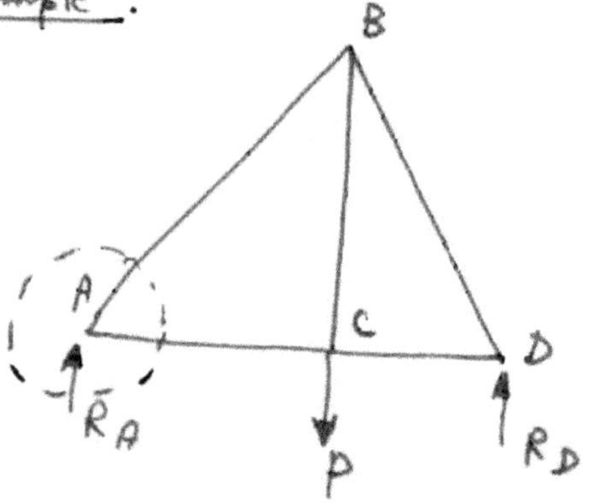

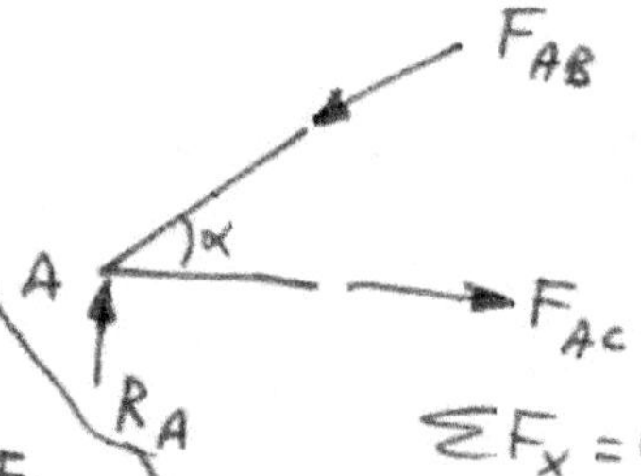

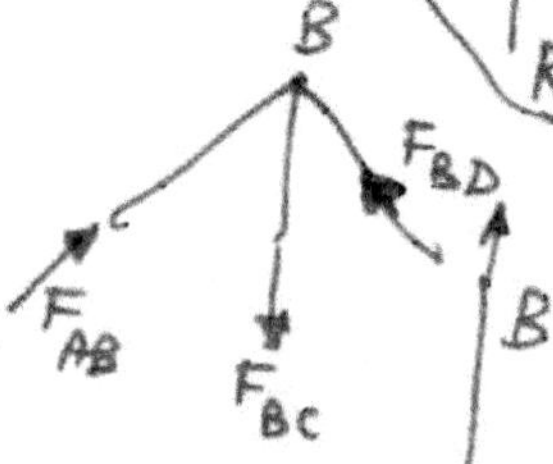

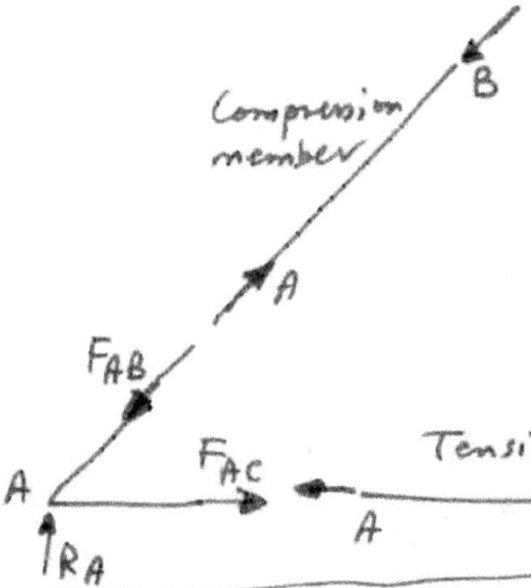

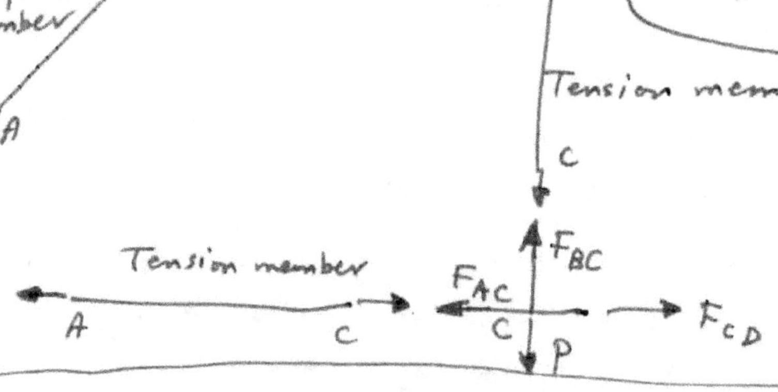

Example 3.1 :

Calculate the bar forces for the truss shown using the method of joints.

Solution :

* The truss is determinate.

(1) Calculate the reactions:

$\Sigma F_x = 0 \Rightarrow R_{Ax} = 0$.

$\Sigma M_A = 0 \;\curvearrowright :$

$R_D(60) - 12(20) - 24(40) = 0$

$\Rightarrow R_D = 20^k \uparrow$

$\Sigma M_D = 0 \;\curvearrowright :$

$-R_{Ay}(60) + 12(40) + 24(20) = 0 \Rightarrow R_{Ay} = 16^k \uparrow$

Check : $\Sigma F_y = 0 \Rightarrow R_{Ay} + R_D - 12 - 24 = 0$

$16 + 20 - 12 - 24 = 0$

$0 = 0 \quad \checkmark$

(2) Apply the method of joints for each joint : $\sqrt{(15)^2 + (20)^2} = 25$.

Joint A : Assume all unknown forces to be tension forces.

$\Sigma F_y = 0 \Rightarrow 16 + F_{AB}\left(\frac{15}{25}\right) = 0$

$\Rightarrow F_{AB} = -26.7^k$.

$\Rightarrow F_{AB} = 26.7^k \; (C)$.

$\Sigma F_x = 0 \Rightarrow F_{AF} + F_{AB}\left(\frac{20}{25}\right) = 0$

$F_{AF} - 26.7\left(\frac{20}{25}\right) = 0 \Rightarrow F_{AF} = 21.4^k \; (T)$.

Joint F :

$\Sigma F_y = 0 \Rightarrow -12 + F_{FB} = 0$

$\Rightarrow F_{FB} = 12^k \; (T)$.

$\Sigma F_x = 0 \Rightarrow F_{FE} - 21.4 = 0$

$\Rightarrow F_{FE} = 21.4 \; (T)$.

Joint B :

$$\Sigma F_y = 0 \Rightarrow 26.7\left(\frac{15}{25}\right) - 12 - F_{BE}\left(\frac{15}{25}\right) = 0$$

$$\Rightarrow F_{BE} = 6.7 \text{ k } (T).$$

$$\Sigma F_x = 0 \Rightarrow F_{BC} + 6.7\left(\frac{20}{25}\right) + 26.7\left(\frac{20}{25}\right) = 0$$

$$\Rightarrow F_{BC} = -26.7 \text{ k}$$

$$\Rightarrow F_{BC} = 26.7^k (C).$$

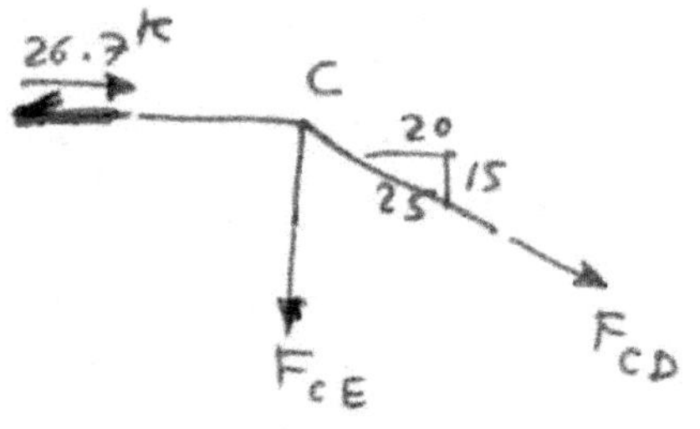

Joint C :

$$\Sigma F_x = 0 \Rightarrow 26.7 + F_{CD}\left(\frac{20}{25}\right) = 0$$

$$\Rightarrow F_{CD} = -33.4 \text{ k}.$$

$$\therefore \quad F_{CD} = 33.4^k (C).$$

$$\Sigma F_y = 0 \Rightarrow -F_{CE} - (-33.4)\left(\frac{15}{25}\right) = 0 \Rightarrow F_{CE} = 20.0 \text{ K}(T).$$

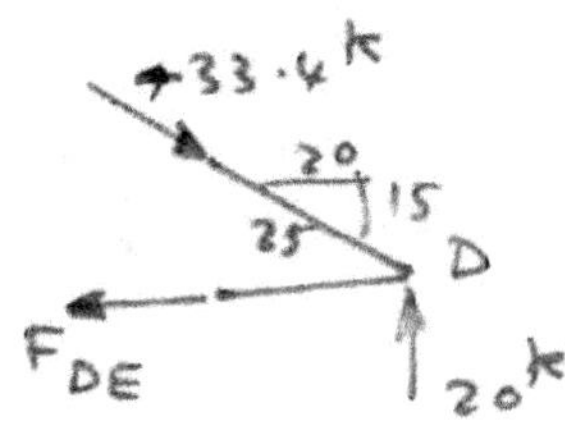

Joint D :

$$\Sigma F_x = 0 \Rightarrow -F_{DE} + 33.4\left(\frac{20}{25}\right) = 0$$

$$\Rightarrow F_{DE} = 26.7^K (T).$$

Check the results :

At joint D : $\Sigma F_y = 0 \Rightarrow 20 - 33.4\left(\frac{15}{25}\right) = 0$

$$\Rightarrow -0.04 = 0 \checkmark$$

At joint E :

$$\Sigma F_x = 0 \Rightarrow 26.7 - 21.4 - 6.7\left(\frac{20}{25}\right) = 0$$

$$\Rightarrow -0.06 = 0 \checkmark$$

$$\Sigma F_y = 0 \Rightarrow 20 - 24 + 6.7\left(\frac{15}{25}\right) = 0$$

$$\Rightarrow 0.02 = 0 \checkmark$$

(3) Record all the bar forces on a diagram of the truss.

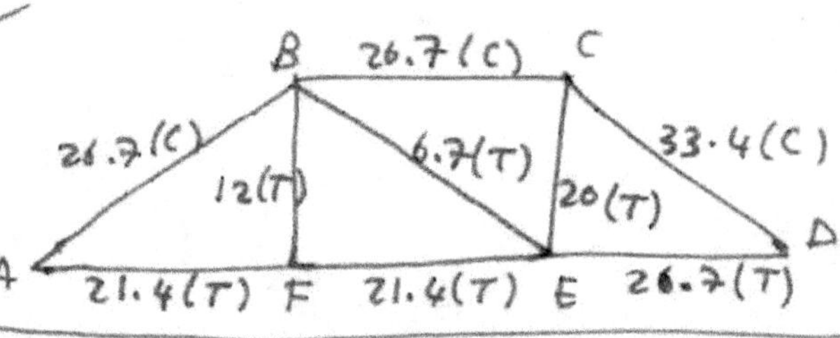

3.5 Method of Sections :

* A section is passed completely through the truss, dividing it into two free bodies.

Example 3.2 : Using the method of sections, determine the forces in bars BC, HG and BG of the truss in the figure.

Solution :

(1) Calculate the reactions :

$\Sigma F_x = 0 \Rightarrow R_{Ax} = 0$.

$\Sigma M_A = 0 \ \circlearrowleft$:

$R_E (32) - 40 (8)$
$- 40 (16) - 40 (24) = 0$
$\Rightarrow R_E = 60 \ kN$.

$\Sigma F_y = 0 \Rightarrow R_{Ay} + 60 - 40 - 40 - 40 = 0 \Rightarrow R_{Ay} = 60 \ kN$.

Alternative method : From Symmetry,

$$R_{Ay} = R_E = \frac{40 + 40 + 40}{2} = 60 \ kN$$.

$\sqrt{(8)^2 + (6)^2} = 10$

(2) Apply the method of sections :

$\Sigma M_G = 0 \ \circlearrowleft$:

$- F_{BC} (6) - 60 (16) + 40 (8) = 0$

$\Rightarrow F_{BC} = -106.7 \ kN$.

$\therefore F_{BC} = 106.7 \ kN \ (C)$.

$\Sigma M_B = 0 \ \circlearrowleft$:

$F_{HG} (6) - 60 (8) = 0 \Rightarrow F_{HG} = 80 \ kN \ (T)$.

$\Sigma F_y = 0 \Rightarrow -F_{BG} \left(\frac{6}{10}\right) - 40 + 60 = 0 \Rightarrow F_{BG} = 33.3 \ kN \ (T)$.

(3) Check : $\Sigma F_x = 0 \Rightarrow F_{BC} + F_{HG} + F_{BG} \left(\frac{8}{10}\right) = 0$

$-106.7 + 80 + 33.3 \left(\frac{8}{10}\right) = 0$

$-0.06 = 0 \ \checkmark$

Example 3.3: Using the method of sections, determine the forces in bars BC, BG and HG for the truss in the figure.

Solution:

(1) Calculate the reactions:

$$\Sigma F_x = 0 \Rightarrow R_{Ax} = 0$$

$$\Sigma M_A = 0 \quad \text{(+)}$$

$$R_E(80) - 20(20) - 30(40) = 0$$
$$\Rightarrow R_E = 20\,k \uparrow$$

$$\Sigma F_y = 0 \Rightarrow R_{Ay} + 20 - 20 - 30 = 0 \Rightarrow R_{Ay} = 30\,k \uparrow$$

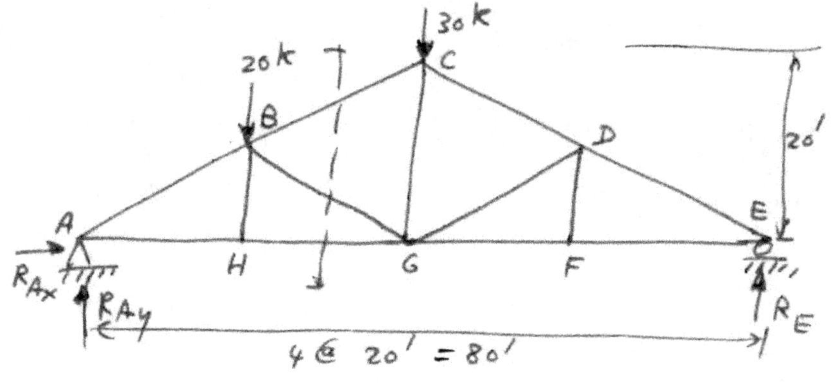

(2) Apply the method of sections:

$$\sqrt{(20)^2 + (10)^2} = 22.36$$

$$\Sigma M_B = 0 \Rightarrow +F_{HG}(10) - 30(20) = 0$$
$$\Rightarrow F_{HG} = 60\,k \quad (T).$$

$$\Sigma M_G = 0 \Rightarrow$$
$$-30(40) + 20(20) - F_{BC}\left(\frac{20}{22.36}\right)(20) = 0$$
$$\Rightarrow F_{BC} = -44.7\,k \Rightarrow F_{BC} = 44.7\,k \quad (C).$$

$$\Sigma M_A = 0 \Rightarrow -20(20) - F_{BG}\left(\frac{10}{22.36}\right)(40) = 0 \Rightarrow F_{BG} = -22.4\,k$$
$$\Rightarrow F_{BG} = 22.4\,k \quad (C).$$

(3) Check the results:

$$\Sigma F_x = 0 \Rightarrow F_{HG} + F_{BC}\left(\frac{20}{22.36}\right) + F_{BG}\left(\frac{20}{22.36}\right) = 0$$
$$60 + (-44.7)\left(\frac{20}{22.36}\right) + (-22.4)\left(\frac{20}{22.36}\right) = 0$$
$$-0.02 = 0 \checkmark$$

$$\Sigma F_y = 0 \Rightarrow 30 - 20 + F_{BC}\left(\frac{10}{22.36}\right) - F_{BG}\left(\frac{10}{22.36}\right) = 0$$
$$10 + (-44.7)\left(\frac{10}{22.36}\right) - (-22.4)\left(\frac{10}{22.36}\right) = 0$$
$$0.03 = 0 \checkmark$$

3.6 Methods of Joints and Sections Combined :

__Example 3.4__ : Determine all the bar forces for the truss shown.

__Solution__ : $\sqrt{(12)^2 + (6)^2} = 13.42$

(1) Calculate the reactions :

$$\Sigma F_x = 0$$
$$\Rightarrow R_{Ax} + 30 = 0$$
$$\Rightarrow R_{Ax} = -30 \text{ kN} \leftarrow$$

$$\Sigma m_A = 0 \quad \circlearrowright \Rightarrow$$

$$R_E(24) - 30(12)$$
$$- 20(6) - 40(12) - 20(18) = 0$$
$$\Rightarrow R_E = 55 \text{ kN} \uparrow$$

$$\Sigma F_y = 0 \Rightarrow R_{Ay} + 55 - 20 - 40 - 20 = 0 \Rightarrow R_{Ay} = 25 \text{ kN} \uparrow$$

(2) Apply both the method of joints and the method of sections :

__* Joint A :__

$$\Sigma F_x = 0 \Rightarrow F_{AF}\left(\frac{12}{13.42}\right) + F_{AB}\left(\frac{6}{13.42}\right) - 30 = 0 \quad (1)$$

$$\Sigma F_y = 0 \Rightarrow 25 + F_{AF}\left(\frac{6}{13.42}\right) + F_{AB}\left(\frac{12}{13.42}\right) = 0 \quad (2)$$

Solve the two equations simultaneously for F_{AB} and F_{AF} :

From eq.(1): $F_{AE} = \left(\frac{13.42}{12}\right)\left[30 - \frac{6}{13.42} F_{AB}\right]$

Substitute in eq.(2) :

$$25 + \left(\frac{6}{13.42}\right)\left(\frac{13.42}{12}\right)\left[30 - \frac{6}{13.42} F_{AB}\right] + \frac{12}{13.42} F_{AB} = 0$$

$$25 + 15 - \frac{3}{13.42} F_{AB} + \frac{12}{13.42} F_{AB} = 0$$

$$40 + \frac{9}{13.42} F_{AB} = 0$$

$$\Rightarrow F_{AB} = -59.6 \text{ kN} \Rightarrow F_{AB} = 59.6 \text{ kN (C)} .$$

$$F_{AF} = \frac{13.42}{12}\left[30 - \frac{6}{13.42}(-59.6)\right] = 63.1 \text{ kN (T)} .$$

* <u>Consider section a-a :</u>

$\sum M_F = 0 \;\circlearrowright\;$: $\sqrt{(6)^2+(3)^2} = 6.71$

$-F_{BC}\left(\frac{6}{6.71}\right)(9) - 30(6) + 20(6)$

$\qquad -30(6) - 25(12) = 0$

$\Rightarrow \; F_{BC} = -67.1 \, kN$

$\qquad \therefore \; F_{BC} = 67.1 \, kN \; (C).$

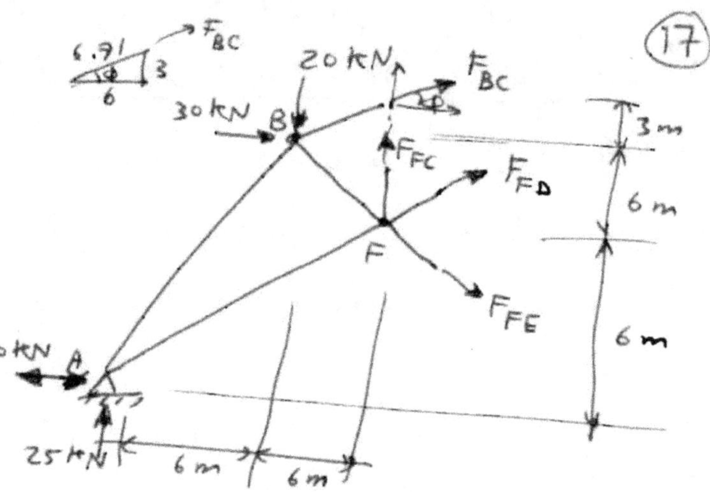

* <u>Joint C :</u>

$\sum F_x = 0 : F_{CD}\left(\frac{6}{6.71}\right) + 67.1\left(\frac{6}{6.71}\right) = 0$

$\Rightarrow \; F_{CD} = -67.1 \, kN$

$\qquad \therefore \; F_{CD} = 67.1 \, kN \; (C).$

$\sum F_y = 0 : -40 - F_{CF} + 67.1\left(\frac{3}{6.71}\right) - (-67.1)\left(\frac{3}{6.71}\right) = 0$

$\qquad \Rightarrow \; F_{CF} = 20.0 \, kN \; (T).$

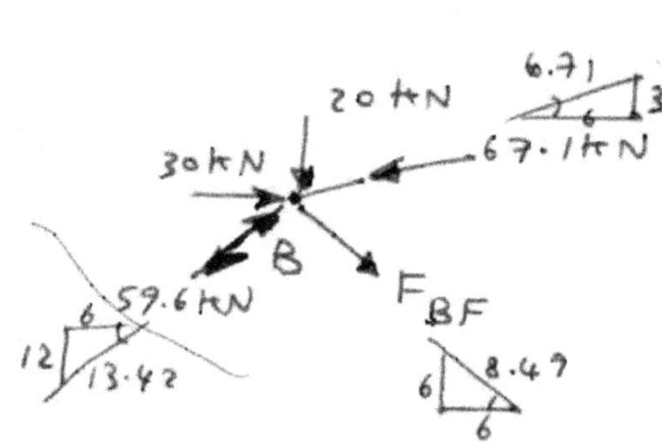

* <u>Joint B :</u>

$\sum F_x = 0 : 30 - 67.1\left(\frac{6}{6.71}\right) + 59.6\left(\frac{6}{13.42}\right)$

$\qquad + F_{BF}\left(\frac{6}{8.49}\right) = 0$

$\qquad \Rightarrow \; F_{BF} = 4.74 \, kN \; (T).$

<u>Check:</u> $\sum F_y = 0 :$

$\quad -20 - 67.1\left(\frac{3}{6.71}\right) + 59.6\left(\frac{12}{13.42}\right) - 4.74\left(\frac{6}{8.49}\right) = 0$

$\qquad\qquad -0.06 = 0 \; \smile$

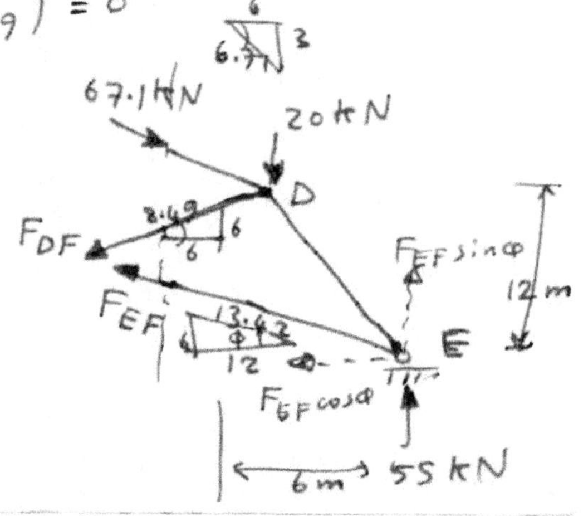

* <u>Consider section b-b :</u>

$\sum M_D = 0 \;\circlearrowright$

$55(6) + F_{EF}\left(\frac{6}{13.42}\right)(6) - F_{EF}\left(\frac{12}{13.42}\right)(12) = 0$

$\left(\frac{36}{13.42} - \frac{144}{13.42}\right) F_{EF} = -55(6)$

$\qquad \Rightarrow \; F_{EF} = 41.0 \, kN \; (T).$

* <u>Joint E :</u>

$\Sigma F_x = 0 : - F_{ED}\left(\dfrac{6}{13.42}\right) - 41.0\left(\dfrac{12}{13.42}\right) = 0$

$\Rightarrow F_{ED} = -82.0 \text{ kN}$

$\therefore F_{ED} = 82.0 \text{ kN} \quad (C).$

Check : $\Sigma F_y = 0 : \quad 55 + (-82.0)\left(\dfrac{12}{13.42}\right) + (41.0)\left(\dfrac{6}{13.42}\right) = 0$

$$0.007 = 0 \checkmark$$

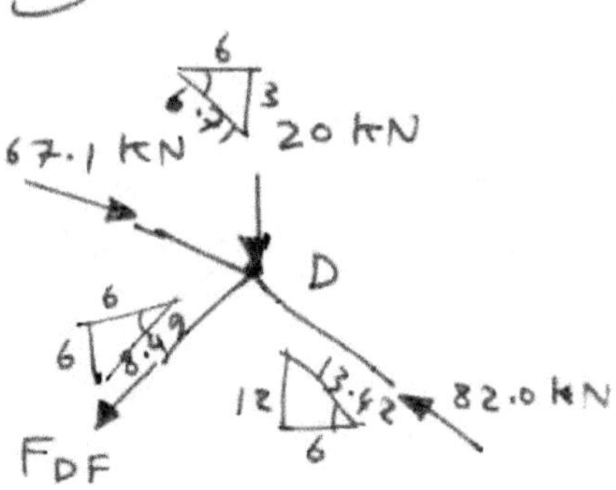

* <u>Joint D :</u>

$\Sigma F_x = 0 :$

$67.1\left(\dfrac{6}{6.71}\right) - F_{DF}\left(\dfrac{6}{8.49}\right) - 82.0\left(\dfrac{6}{13.42}\right) = 0$

$\Rightarrow F_{DF} = 33.0 \text{ kN} \quad (T).$

Check : $\Sigma F_y = 0 :$

$-20 - 67.1\left(\dfrac{3}{6.71}\right) - 33.0\left(\dfrac{6}{8.49}\right) + 82.0\left(\dfrac{12}{13.42}\right) = 0$

$$0.002 = 0 \checkmark$$

<u>Final Check at Joint F :</u>

$\Sigma F_x = 0 :$

$33\left(\dfrac{6}{8.49}\right) - 4.74\left(\dfrac{6}{8.49}\right)$

$+ 41\left(\dfrac{12}{13.42}\right) - 63.1\left(\dfrac{12}{13.42}\right) = 0$

$\Rightarrow \quad 0.2 = 0 \checkmark$

$\Sigma F_y = 0 :$

$20 + 33\left(\dfrac{6}{8.49}\right) + 4.74\left(\dfrac{6}{8.49}\right) - 41\left(\dfrac{6}{13.42}\right) - 63.1\left(\dfrac{6}{13.42}\right) = 0$

$\Rightarrow \quad 0.1 = 0 \checkmark$

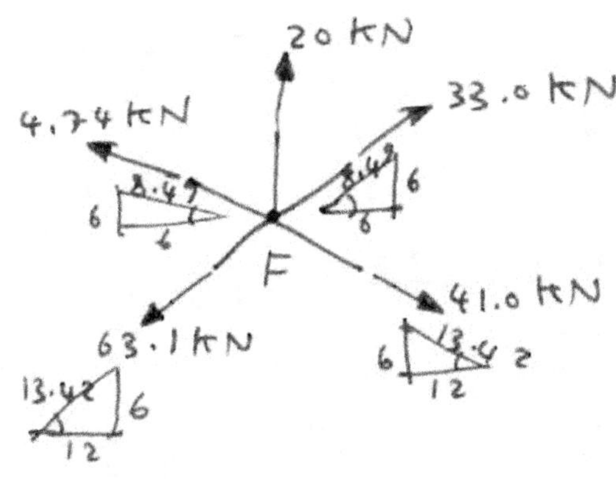

Example :

Find the force in member AB of the Baltimore truss shown.

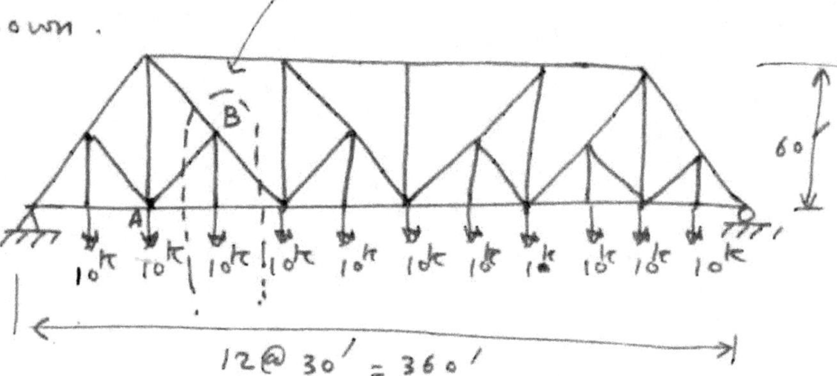

Solution :

Consider the section shown :

$\overset{+}{\curvearrowleft} \Sigma M_0 = 0$:

$$F_{AB}\left(\frac{30}{42.4}\right)(60) + 10(30) = 0$$

$$\Rightarrow F_{AB} = -7.07\,k$$

$$\therefore F_{AB} = 7.07\,k\ (C).$$

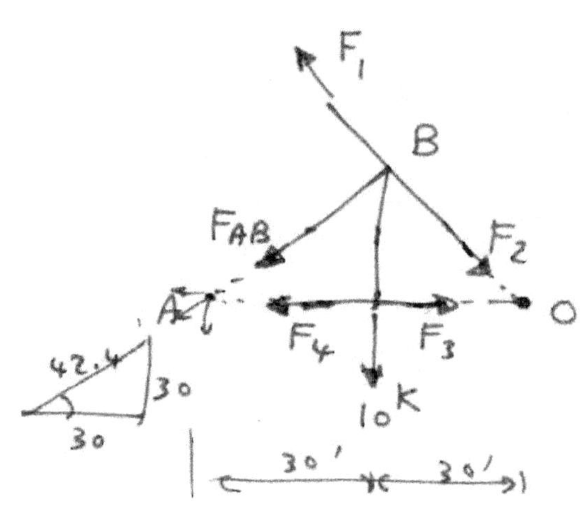

Example :

Find the force in member AB of the K-truss shown.

Solution :

Consider the section shown :

First, find the reactions.

From symmetry,

$$R_1 = R_2 = \frac{7(20)}{2} = 70\,k.$$

$\Sigma M_0 = 0$:

$$-F_{AB}(40) - 70(80) + 20(40) = 0$$

$$\Rightarrow F_{AB} = -120\,k.$$

$$\therefore F_{AB} = 120\,k\ (C).$$

Example :

Find the force in
member EB ?

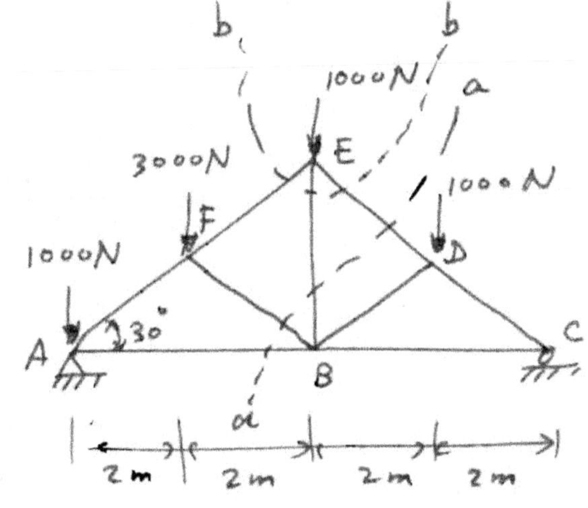

Solution :

(1) Find the reactions.

(2) Section (a–a):

$$F_{ED} = -3000 N .$$

(3) Section (b–b) — Joint E

$$F_{EF} = -3000 N$$
$$F_{EB} = 2000 N (T) .$$

Show the bar forces on a diagram of the truss :

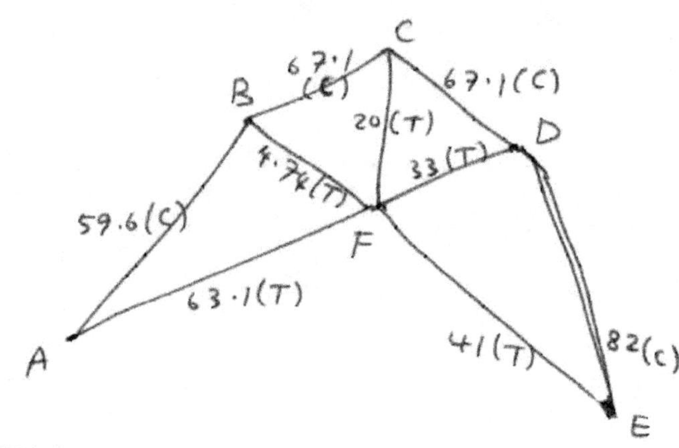

3.7 Stability and Determinacy of Trusses :

* A <u>determinate</u> structure is one for which the number of equations of equilibrium is equal to the number of unknown reactions.

* An <u>indeterminate</u> structure is one that possesses more unknown reactions than equations of equilibrium.

* An <u>unstable</u> structure is one that has an insufficient number of reactions to prevent motion.

For trusses, the stability and determinacy depends <u>not</u> only on the number of reactions, but also on the number and arrangement of individual members.

* A truss is <u>externally stable and determinate</u> if it possesses the proper number of reactions.

* A truss is <u>internally stable and determinate</u> if it has the correct number of bars arranged in a proper manner.

* At each joint, we write <u>two</u> independent equations of equilibrium.
    $$\Sigma F_x = 0 \quad \text{and} \quad \Sigma F_y = 0 .$$

If a truss has j joints $\Longrightarrow$ $2j$ equations of equilibrium.

The unknowns are the number of reactions $= r$

the number of members $= m$

* A truss is <u>determinate</u> if the number of unknowns (the number of member forces and reactions) is equal to the number of equations (twice the number of joints), i.e.

$$\boxed{m + r = 2j} \qquad\qquad (1)$$

* A truss is <u>indeterminate</u> if $\quad m + r > 2j$

* A truss is <u>unstable</u> if $\quad m + r < 2j$

<u>Note</u> : The condition given by equation (1) is a <u>necessary</u> but <u>insufficient</u> condition for stability and determinacy.

* In addition to satisfying equation (1), a stable and determinate truss must have both the correct number of reactions and the correct number of bars arranged in a proper manner.

<u>Example (1):</u>

$$j = 4$$
$$r = 3$$
$$m = 5$$
$$\therefore \; 2j = m + r$$
$$2(4) = 5 + 3$$
$$8 = 8 \;\checkmark$$

$\therefore$ the truss is <u>stable</u> and <u>determinate</u>.

(1) The truss is <u>externally</u> stable and determinate because it has three nonparallel, nonconcurrent reactions that suffice to prevent motion of the structure, and that can be evaluated using the three equations of equilibrium.

(2) The truss is <u>internally</u> stable and determinate because the bars are arranged to form one or more triangles and the bar forces can be evaluated using the equations of equilibrium.

<u>Example (2):</u>

$$j = 4$$
$$r = 3$$
$$m = 4$$
$$2j = 2(4) = 8$$
$$m + r = 4 + 3 = 7$$
$$\therefore \; 2j > m + r$$

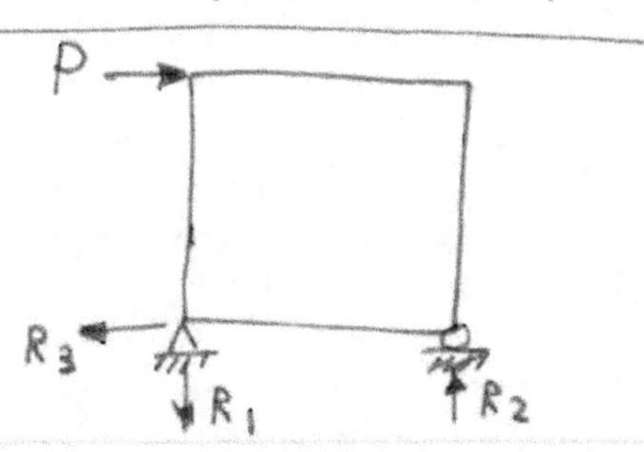

$\therefore$ the truss is internally <u>unstable</u>.

   The rectangle can collapse.

(However, the truss is externally stable and determinate).

  $\therefore$ the truss is <u>unstable</u>.

Example (3) :

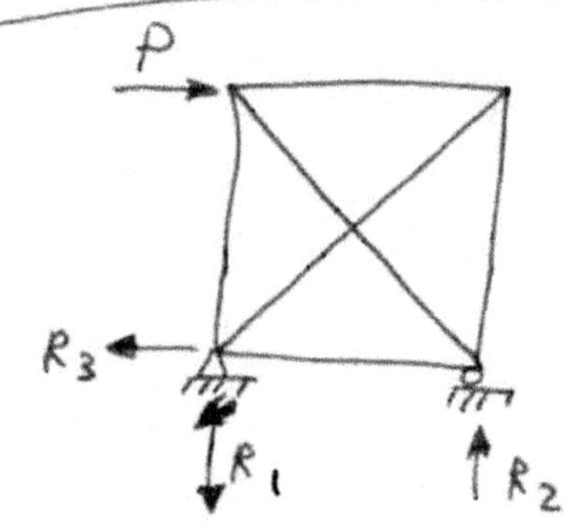

$$j = 4$$
$$r = 3$$
$$m = 6$$
$$2j = 2(4) = 8$$
$$m + r = 6 + 3 = 9$$
$$2j < m + r .$$

$\therefore$ The truss is internally <u>indeterminate</u>.

  The bar forces <u>cannot</u> be evaluated using <u>only</u> the equations of equilibrium.

   We cannot use the method of joints or the method of sections (since each joint has three unknowns but only two equations of equilibrium).

Note that the reactions can be evaluated using the equations of equilibrium $\Longrightarrow$ the truss is <u>externally determinate</u>.

  $\therefore$ The truss is <u>indeterminate</u> (but <u>stable</u>).

Example (4) :

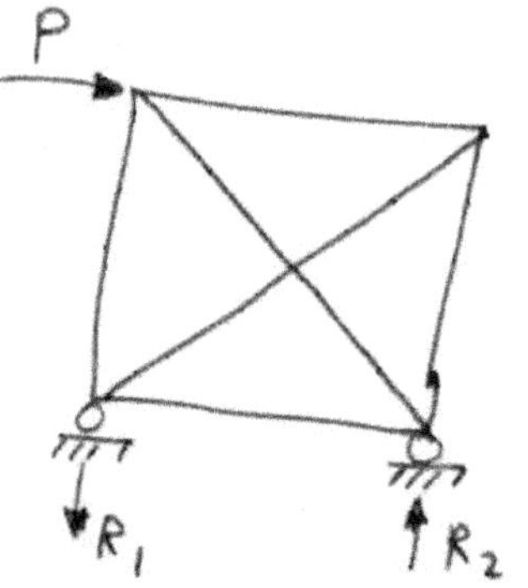

$$j = 4$$
$$r = 2$$
$$m = 6$$
$$2j = 2(4) = 8$$
$$m + r = 6 + 2 = 8$$
$$\therefore \; 2j = m + r$$

However, this is an <u>insufficient</u> condition.

* The truss is <u>externally unstable</u> because the two vertical reactions <u>cannot</u> prevent motion of the structure.

* The truss is <u>internally indeterminate</u> because it has one more diagonal than necessary.

  $\therefore$ the truss is <u>unstable</u> and indeterminate.

Graphical Solutions : Maxwell Diagram (not required).

Types of Trusses :

(1) Simple Trusses :

Simple trusses are formed by constructing triangles by adding two members and a joint.

Example :

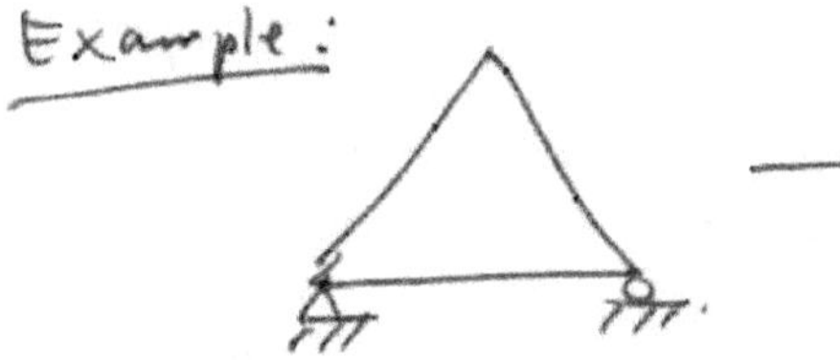 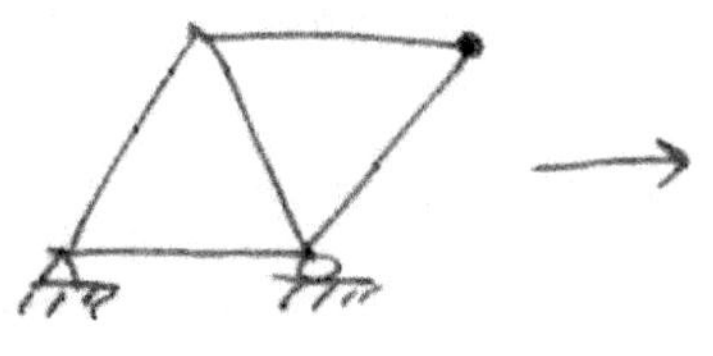 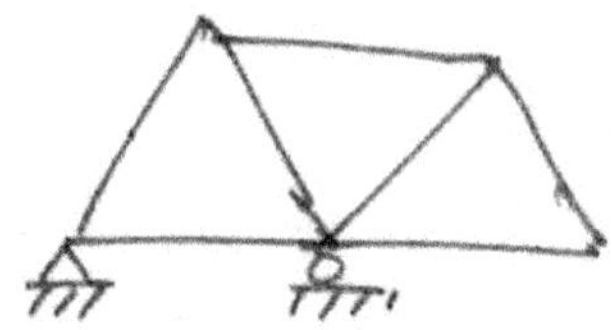

(2) Compound Trusses :

A compound truss is a truss made by connecting two or more simple trusses.

The simple trusses may be connected by one joint and one link, by a connecting truss, by two or more joints, and so on.

Example :

$j = 15$

$m = 27$

$r = 3$

$2j = 30$

$m + r = 30$ $\left.\right\}$ $\therefore 2j = m + r$

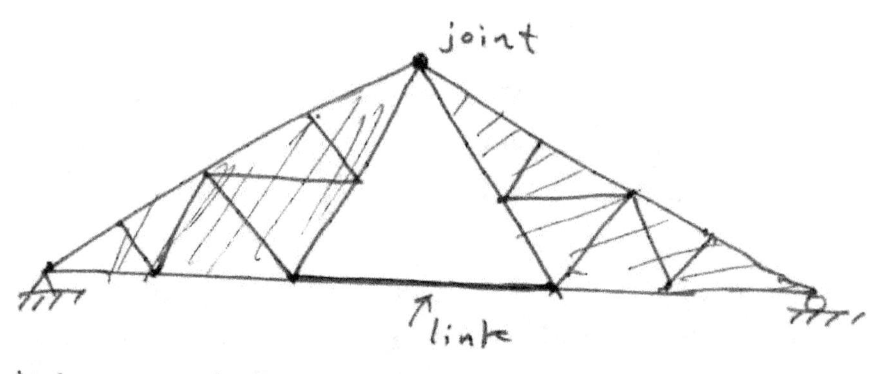

$\therefore$ the truss is determinate.

(3) Complex Trusses :

A complex truss is a truss that is neither a simple truss nor a compound truss.

It is characterized by three members at each joint.

Example :

$j = 6$

$m = 9$

$r = 3$

$2j = 12$

$m + r = 12$ $\left.\right\}$ $2j = m + r$

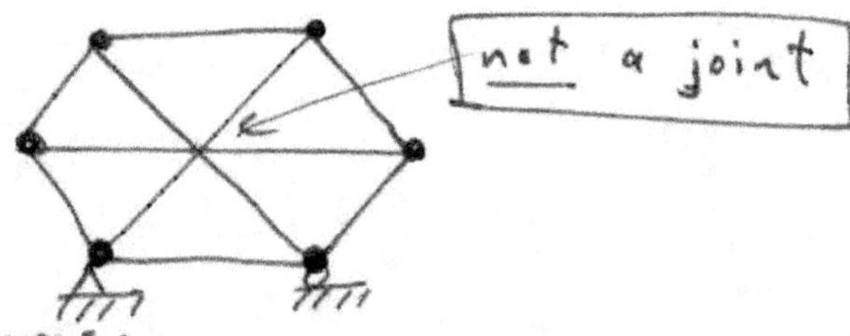

Solve 12 equations in 12 unknowns.

* A truss is <u>stable</u> and determinate if it satisfies $2j = m + r$ and is stable and determinate both <u>internally</u> and <u>externally</u>.

3.8 Computer Solution:

* There are two ways of analyzing a determinate truss using the computer:

(1) <u>Method 1</u>: We write the equations of equilibrium at each joint and use the computer to solve the set of simultaneous equations for the reactions and bar forces.

(2) <u>Method 2</u>: We use the computer to set up the equations and solve them. (The program is already written – pages <u>(63–65)</u>.

Example:

Find all the bar forces for the truss shown.

<u>Solution (1)</u>:

There are <u>six</u> unknowns:

   (1) three reactions

   (2) three bar forces.

∴ We need <u>six</u> equations (two equations at each joint).

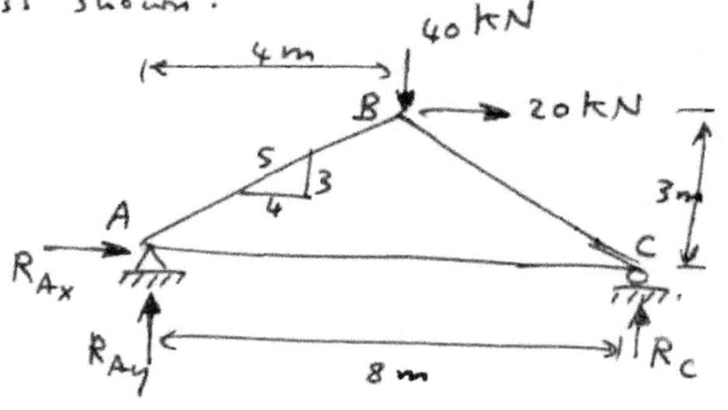

Joint A:

$\Sigma F_x = 0$: $R_{Ax} + F_{AC} + F_{AB}\left(\frac{4}{5}\right) = 0$ —— (1)

$\Sigma F_y = 0$: $R_{Ay} + F_{AB}\left(\frac{3}{5}\right) = 0$ —— (2)

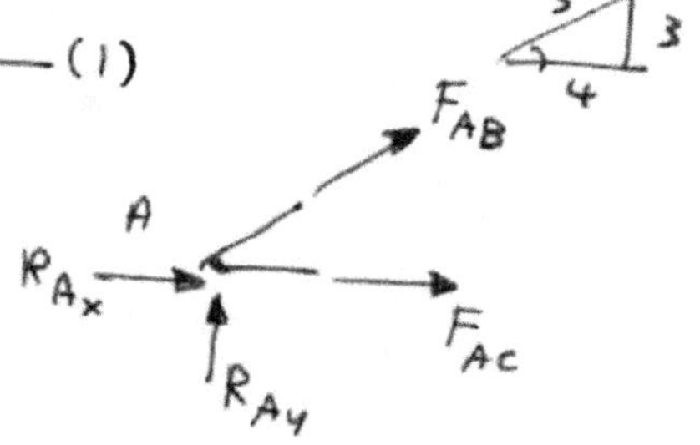

Joint B:

$\Sigma F_x = 0$:

$20 + F_{BC}\left(\frac{4}{5}\right) - F_{AB}\left(\frac{4}{5}\right) = 0$ —— (3)

$\Sigma F_y = 0$:

$-40 - F_{AB}\left(\frac{3}{5}\right) - F_{BC}\left(\frac{3}{5}\right) = 0$ —— (4)

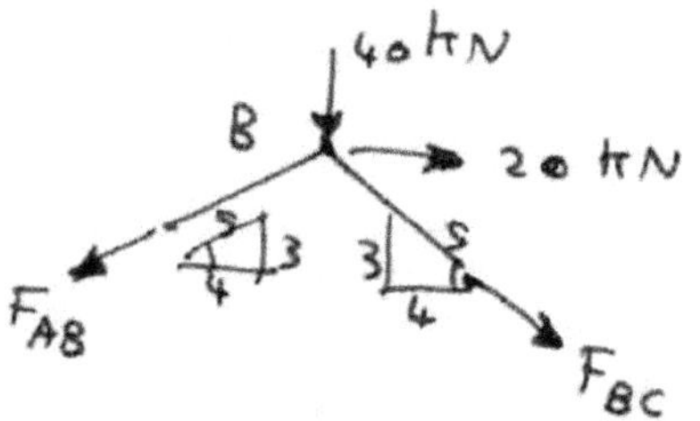

$$
\begin{array}{cccccc}
R_{Ax} & R_{Ay} & R_C & F_{AB} & F_{AC} & F_{BC}
\end{array}
$$

$$
\begin{bmatrix}
1 & 0 & 0 & \frac{4}{5} & 1 & 0 \\
0 & 1 & 0 & \frac{3}{5} & 0 & 0 \\
0 & 0 & 0 & -\frac{4}{5} & 0 & \frac{4}{5} \\
0 & 0 & 0 & -\frac{3}{5} & 0 & -\frac{3}{5} \\
0 & 0 & 0 & 0 & -1 & -\frac{4}{3} \\
0 & 0 & 1 & 0 & 0 & \frac{3}{5}
\end{bmatrix}
\begin{Bmatrix}
R_{Ax} \\ R_{Ay} \\ R_C \\ F_{AB} \\ F_{AC} \\ F_{BC}
\end{Bmatrix}
=
\begin{Bmatrix}
0 \\ 0 \\ -20 \\ 40 \\ 0 \\ 0
\end{Bmatrix}
$$

Solve the system using program __SOLVE__ on the computer:

The results are:

$$R_{Ax} = -2.0 \times 10^1 = -20 \ kN$$

$$R_{Ay} = 1.25 \times 10^1 = 12.5 \ kN$$

$$R_C = 2.75 \times 10^1 = 27.5 \ kN$$

$$F_{AB} = -2.0833 \times 10^1 = -20.83 \ kN$$

$$F_{AC} = 3.667 \times 10^1 = 36.67 \ kN$$

$$F_{BC} = -4.583 \times 10^1 = -45.83 \ kN$$

Input Data:

(1) Enter the number of equations:
    Enter 6

(2) Enter coefficients of equation 1
    1, 0, 0, 0.8, 1, 0, 0
    Enter coefficients of equation 2
    0, 1, 0, 0.6, 0, 0, 0
    Enter coefficients of equation 3
    0, 0, 0, -0.8, 0, 0.8, -20
    Enter coefficients of equation 4
    0, 0, 0, -0.6, 0, -0.6, 40

Enter linear system one equation at a time. Conclude each equation with its right-hand constant:

(continued on)

```fortran
      PROGRAM SOLVE
C This program will solve a linear system A*X=B
      DIMENSION A(50,50),B(50)
      PRINT*, ' Enter the number of equations:'
      READ*, N
      IF(N.EQ.0) STOP
      PRINT*, 'Enter the linear system one equation at a time:'
      PRINT*, 'Conclude each equation with its right-hand constant:'
      DO 20 I=1,N
        PRINT*, 'Enter coefficients of equation', I
   20 READ*, (A(I,J),J=1,N), B(I)
C Solve the linear system
      CALL LINSYS(A,B,N,50,IER)
      PRINT 1000, N, IER
 1000 FORMAT(///, ' N=',I2,5X,'IER=',I3,//,'    I          Solution',/)
      PRINT 1001, (I,B(I),I=1,N)
 1001 FORMAT(I3,1PE20.10)
      STOP
      END
C
      SUBROUTINE LINSYS(MAT,B,N,MD,IER)
C This subroutine solves a system of linear equations
C             A*X=B
C The method used is Gaussian elimination with partial pivoting
C Input:
C The coefficient matrix A is stored in the array MAT
C The right-side constants are stored in the array B
C The order of the linear system is N (number of equations)
C The variable MD is the number of rows that MAT is dimensioned
C as having in the calling program
C Output:
C The array B contains the solution X
C MAT contains the upper triangular matrix U obtained by elimination
C The row multipliers used in the elimination are stored in
C the lower triangular part of MAT
C IER=0 means the matrix A was computationally nonsingular
C and the Gaussian elimination was completed satisfactorily.
C IER=1 means that the matrix A was computationally singular
      INTEGER PIVOT
      REAL MULT,MAT
      DIMENSION MAT(MD,*),B(*),PIVOT(100)
C Begin eliminations steps
      DO 40 K=1,N-1
C Choose pivot row
      PIVOT(K)=K
      AMAX=ABS(MAT(K,K))
      DO 10 I=K+1,N
        ABSA=ABS(MAT(I,K))
        IF(ABSA.GT.AMAX) THEN
          PIVOT(K)=I
          AMAX=ABSA
        END IF
   10   CONTINUE
      IF(AMAX.EQ.0.0) THEN
C Coefficient matrix is singular
        IER=1
        RETURN
      END IF
      IF(PIVOT(K).NE.K) THEN
C Switch rows K and PIVOT(K)
        I=PIVOT(K)
        TEMP=B(K)
        B(K)=B(I)
        B(I)=TEMP
        DO 20 J=K,N
          TEMP=MAT(K,J)
          MAT(K,J)=MAT(I,J)
   20     MAT(I,J)=TEMP
      END IF
C Perform step #K of elimination
      DO 30 I=K+1,N
        MULT=MAT(I,K)/MAT(K,K)
        MAT(I,K)=MULT
        B(I)=B(I)-MULT*B(K)
        DO 30 J=K+1,N
   30   MAT(I,J)=MAT(I,J)-MULT*MAT(K,J)
   40 CONTINUE
      IF(MAT(N,N).EQ.0.0) THEN
C Coefficient matrix is singular
        IER=1
        RETURN
      END IF
C Solve for solution X using back substitution
      DO 60 I=N,1,-1
        SUM=0.0
        DO 50 J=I+1,N
   50   SUM=SUM+MAT(I,J)*B(J)
   60 B(I)=(B(I)-SUM)/MAT(I,I)
      IER=0
      RETURN
      END
```

$\rightarrow$ _continued_

Enter coefficients of equation 5

     0, 0, 0, 0, −1, −0.8, 0

Enter coefficients of equation 6

     0, 0, 1, 0, 0, 0.6, 0

$$\Sigma F_x = 0 : -F_{AC} - F_{BC}\left(\tfrac{4}{5}\right) = 0 \quad\underline{\quad} (5)$$

$$\Sigma F_y = 0 : R_C + F_{BC}\left(\tfrac{3}{5}\right) = 0 \quad\underline{\quad} (6)$$

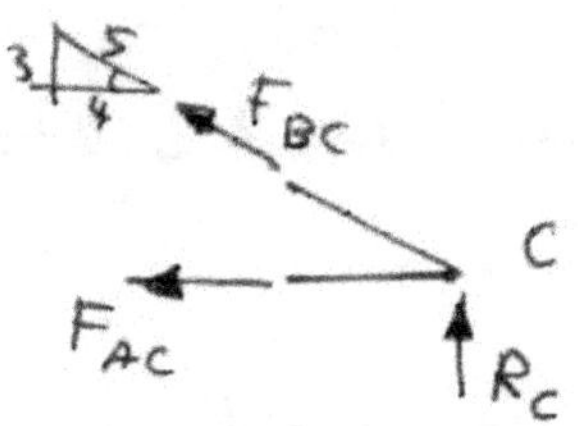

We now use the computer to solve the __six__ equations for the six unknowns (e.g. use Gaussian elimination):

the results are:

$$F_{AB} = -20.83 \text{ kN}, \quad F_{AC} = +36.67 \text{ kN}, \quad F_{BC} = -45.83 \text{ kN}$$

$$R_{Ay} = +12.50 \text{ kN}, \quad R_{Ax} = -20.0 \text{ kN}, \quad R_C = +27.50 \text{ kN}$$

<u>Solution (2)</u>: Use the program on pages __63—65__ to analyze the truss:

Steps:

(1) Choose $x-$ and $y-$directions.

(2) Number the members and the reactions (the unknowns) as shown in the figure. $(F_1, F_2, F_3, F_4, F_5, F_6)$.

e.g. F_2 is the force in member BC.
F_5 is the vertical reaction at A.

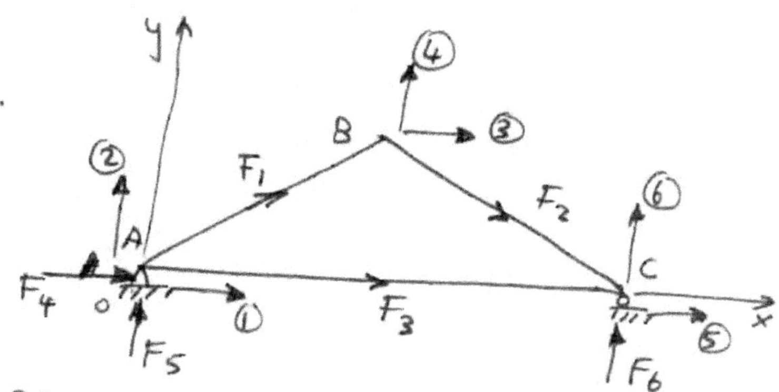

(3) Give each member a direction.

(4) Assume all bars to be in tension, and the reactions along the positive $x-$ and $y-$directions.

(5) Assign two numbered vectors to each joint, one for the $x-$direction and one for the $y-$direction. (__vector numbers__).

(6) Determine the $(x,y)-$coordinates of the joints.
e.g. Coordinates of B are $C(3) = 4m$, $C(4) = 3m$.

The computer will ask a series of questions:
Preparation of __input data__ to the program:
(1) Enter the number of simultaneous equations.

joint	x	y
A	0	0
B	4	3
C	8	0

```fortran
C       -------------------------------------------------
C           FORTRAN 77 Program for Analyzing Trusses
C       -------------------------------------------------
        PROGRAM TRUSS
        DIMENSION A(50,50), P(50), F(50),C(50)
        PRINT*, 'Enter the number of simultaneous equations;'
        READ*,N
        PRINT*, 'Enter the number of members;'
        READ*,NM
        PRINT*, 'Enter the values of applied loads;'
        DO 5 I=1,N
          PRINT*, 'P(',I,'):'
          READ*, P(I)
          P(I)=-P(I)
    5   CONTINUE
        PRINT*, 'Enter the x and y coordinates at each joint'
        DO 10 I=1,N
          PRINT*, 'C(',I,'):'
          READ*, C(I)
   10   CONTINUE
        DO 15 M=1,NM
          PRINT*, 'Enter the vector numbers  i, j, k, l for member ', M
          READ*, I, J, K, L
          ALENGTH= SQRT((((C(K)-C(I))**2) + ((C(L)-C(J))**2))
          A(I,M)=(C(K)-C(I))/ALENGTH
          A(J,M)=(C(L)-C(J))/ALENGTH
          A(K,M)=-A(I,M)
          A(L,M)=-A(J,M)
   15   CONTINUE
        DO 20 M= (NM+1), N
          PRINT*, 'Enter the vector number for reaction ', M
          READ*, I
          A(I,M)=1.0
   20   CONTINUE
        CALL SIMUS(A,P,F,N)
        PRINT*, ' Solution:'
        PRINT*, ' ---------'
        DO 25 I=1,NM
          PRINT 40, ' Member     ', I, F(I)
   25   CONTINUE
        DO 30 I= (NM+1), N
          PRINT 40, ' Reaction   ', I, F(I)
   30   CONTINUE
   40   FORMAT(A ,1X,I2,F12.5)
        END
C
        SUBROUTINE SIMUS(A,B,X,N)
        DIMENSION A(50,50), B(50), X(50), IQ(50)
 1000   FORMAT(/,1X, 'Coefficient matrix is singular')
        DO 10 I=1,N
   10   IQ(I)=I
        NN=N-1
        DO 55 I=1,NN
          II=I+1
          PIVOT=ABS(A(I,I))
          IR=I
          IC=I
          DO 15 J=I,N
            DO 15 K=I,N
              AA=ABS(A(J,K))
              IF(AA.LE.PIVOT) GO TO 15
              PIVOT = AA
              IR=J
              IC=K
   15     CONTINUE
          IF(PIVOT.NE.0) GO TO 20
          PRINT 1000
          GO TO 100
   20     IF(IR.EQ.I) GO TO 30
          DO 25 J=I,N
            AA=A(I,J)
            A(I,J)=A(IR,J)
   25     A(IR,J)=AA
          AA=B(I)
          B(I)=B(IR)
          B(IR)=AA
   30     IF(IC.EQ.I) GO TO 40
          DO 35 J=1,N
            AA=A(J,I)
            A(J,I)=A(J,IC)
   35     A(J,IC)=AA
          K=IQ(I)
          IQ(I)=IQ(IC)
          IQ(IC)=K
   40     DO 50 J=II,N
            C=A(J,I)/A(I,I)
            DO 45 K=II,N
   45       A(J,K)=A(J,K)-C*A(I,K)
   50     B(J)=B(J)-C*B(I)
   55   CONTINUE
        X(N)=B(N)/A(N,N)
        DO 65 I1=1,NN
          I=N-I1
          X(I)=B(I)
          II=I+1
          DO 60 J=II,N
   60     X(I)=X(I)-A(I,J)*X(J)
   65   X(I)=X(I)/A(I,I)
        DO 75 I=1,N
          IF(IQ(I).EQ.I) GO TO 75
          II=I+1
          DO 70 J=II,N
            IF(IQ(J).NE.I) GO TO 70
            AA=X(I)
            X(I)=X(J)
            X(J)=AA
            IQ(J)=IQ(I)
            GO TO 75
   70     CONTINUE
   75   CONTINUE
  100   RETURN
        END
```

This is <u>twice</u> the number of joints.
Enter 6.

(2) Enter the number of members. Enter 3.

(3) Enter the values of applied loads.

$$P(1) = 0$$
$$P(2) = 0$$
$$P(3) = 20$$
$$P(4) = -40$$
$$P(5) = 0$$
$$P(6) = 0$$

(4) Enter the x and y coordinates at each joint.

$$C(1) = 0$$
$$C(2) = 0$$
$$C(3) = 4$$
$$C(4) = 3$$
$$C(5) = 8$$
$$C(6) = 0$$

(5) Enter the vector numbers i, j, k, l for member 1
Enter 1, 2, 3, 4.

Enter the vector numbers i, j, k, l for member 2
Enter 3, 4, 5, 6

Enter the vector numbers i, j, k, l for member 3
Enter 1, 2, 5, 6

(6) Enter the vector number for reaction 4
Enter 1

Enter the vector number for reaction 5
Enter 2

Enter the vector number for reaction 6
Enter 6

<u>Solution</u> :

Member	1	-20.83333
Member	2	-45.83333
Member	3	36.6667
Reaction	4	-20.0000
Reaction	5	12.5000
Reaction	6	27.5000

K-truss, there is <u>no</u> need to find the reactions.

Homework 2

3.2 , 3.8 , 3.12 , 3.16 , 3.20 .

3.24 .

← use the computer program. Prepare all the <u>input data</u>.

TRUSS

<u>Stability and Determinacy of Trusses :</u>

* A truss is <u>indeterminate</u> if

No. of Unknowns > No. of Equations

$$m + r > 2j \qquad \text{(plane trusses)}$$
$$m + r > 3j \qquad \text{(space trusses)}.$$

* Degree of Indeterminacy = No. of Unknowns — No. of equations

* A truss is <u>unstable</u> if :

① No. of Unknowns < No. of Equations

or
② No. of Unknowns $\geq$ No. of Equations and   one or more
of the following conditions is satisfied :

(a) The reactions are parallel
(b) The reactions are concurrent.

$\rightarrow$ <u>externally unstable</u>

$j = 6$
$m = 9$
$r = 3$
$m + r = 2j$
$12 = 12$ ✓

(c) the members are <u>not</u> arranged in a proper manner :
(<u>internally unstable</u>).

$j = 6$
$m = 8$
$r = 4$
$m + r = 2j$
$12 = 12$

a member is
missing

(d) the <u>determinant</u> of the matrix $|D| = 0$.
(externally and/or internally unstable) .
(For complex and space trusses) .

To solve the equations, we must have $|D| \neq 0$.

Chapter 4: Space Trusses

Calculation of Reactions:

$$\Sigma F_x = 0 \qquad \Sigma M_x = 0$$
$$\Sigma F_y = 0 \qquad \Sigma M_y = 0$$
$$\Sigma F_z = 0 \qquad \Sigma M_z = 0$$

Equations of equilibrium in three dimensions. (there are six equations).

∴ We can find six unknown reactions at the supports of a three-dimensional truss (space truss).

* There are three types of supports for a space truss:

(1) A ball and socket joint that can resist translation in three perpendicular directions ⟹ three reactions.

(2) A support that resists movement in only two directions ⟹ two reactions.

(3) A support that prevents motion in only one direction. ⟹ one reaction.

Calculation of Member Forces:

Simplifying Assumptions

(1) The members of a space truss are connected to one another by ball and socket joints.

(2) External loads are applied to the truss only at the joints.

(3) The bars are subjected to axial forces only.

The Method of Joints:

At each joints, we have three equations of equilibrium:

$$\begin{cases} \Sigma F_x = 0 \\ \Sigma F_y = 0 \\ \Sigma F_z = 0 \end{cases}$$

Resolve F_{AB} into its three rectangular components.

$$F_{AB,x} = F_{AB}\cos\theta_x = F_{AB}\left(\frac{AC}{AB}\right)$$

$$F_{AB,y} = F_{AB}\cos\theta_y = F_{AB}\left(\frac{AF}{AB}\right)$$

$$F_{AB,z} = F_{AB}\cos\theta_z = F_{AB}\left(\frac{AE}{AB}\right)$$

where

$$AB = \sqrt{(AC)^2 + (AF)^2 + (AE)^2}$$

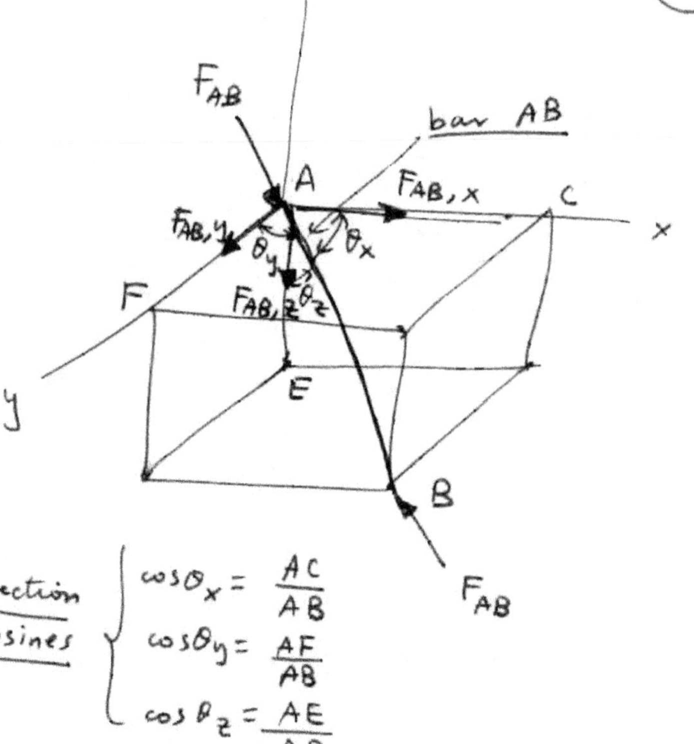

Direction Cosines

$$\cos\theta_x = \frac{AC}{AB}$$
$$\cos\theta_y = \frac{AF}{AB}$$
$$\cos\theta_z = \frac{AE}{AB}$$

Example 4.1:

Determine the reactions and the bar forces for the truss shown in the figure.

Solution:

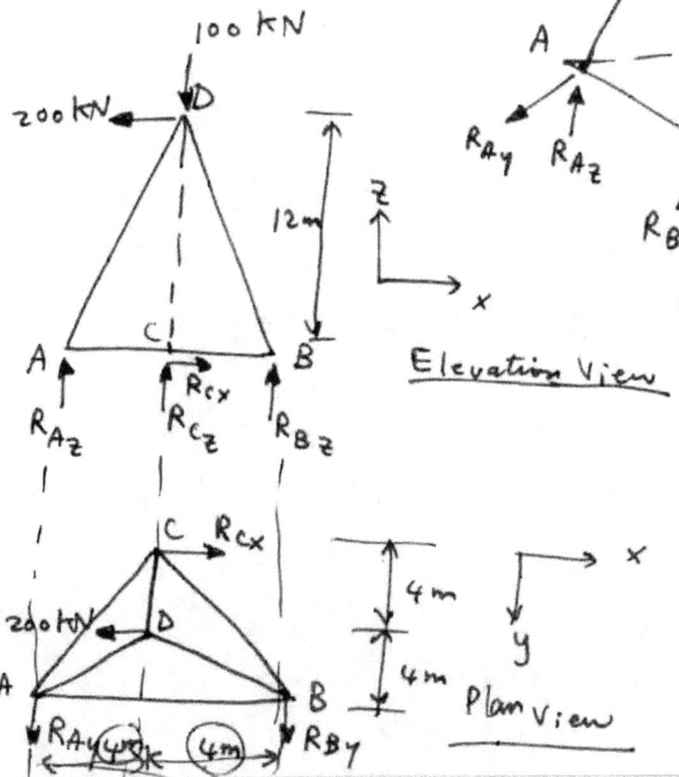

(1) Find the reactions:

$$\sum M_{x(AB)} = 0 : \quad +R_{cz}(8) - 100(4) = 0$$

$$\Rightarrow R_{cz} = 50 \text{ kN} \uparrow$$

$$\sum M_{y(A)} = 0 : \quad R_{Bz}(8) + \overset{=50}{R_{cz}}(4) - 100(4) + 200(12) = 0$$

$$\Rightarrow R_{Bz} = -275 \text{ kN}.$$

$$\therefore R_{Bz} = 275 \text{ kN} \downarrow$$

$$\sum F_z = 0 : \quad R_{Az} - 275 + 50 - 100 = 0$$

$$\Rightarrow R_{Az} = 325 \text{ kN} \uparrow$$

$$\sum F_x = 0 : \quad R_{cx} - 200 = 0 \Rightarrow R_{cx} = 200 \text{ kN} \searrow$$

$$\sum M_{z(B)} = 0 : \quad -\overset{200}{R_{cx}}(8) + R_{Ay}(8) + 200(4) = 0$$

$$\Rightarrow R_{Ay} = 100 \text{ kN} \swarrow$$

$$\sum F_y = 0 : \quad 100 + R_{By} = 0 \Rightarrow R_{By} = -100 \text{ kN}.$$

$$\therefore R_{By} = 100 \text{ kN} \nearrow$$

(2) Use the method of joints:

Member	Projection			Ratio of Projection to length (direction cosines)			Length
	x	y	z	x/L	y/L	z/L	
AB	8	0	0	1.0	0	0	8.0
AC	4	8	0	0.447	0.895	0	8.94
AD	4	4	12	0.301	0.301	0.904	13.27
BC	4	8	0	0.447	0.895	0	8.94
BD	4	4	12	0.301	0.301	0.904	13.27
CD	0	4	12	0	0.316	0.949	12.65

Joint A :

$\Sigma F_x = 0$:

$\quad 0.301\,F_{AD} + 0.447\,F_{AC} + F_{AB} = 0 \quad\text{——(1)}$

$\Sigma F_y = 0$: $\quad 100 - 0.301\,F_{AD} - 0.895\,F_{AC} = 0 \quad\text{——(2)}$

$\Sigma F_z = 0$: $\quad 325 + 0.904\,F_{AD} = 0$

$\quad\Rightarrow\ F_{AD} = \dfrac{-359.5}{}\ kN$

$\qquad\qquad F_{AD} = 359.5\ kN\ (C).$

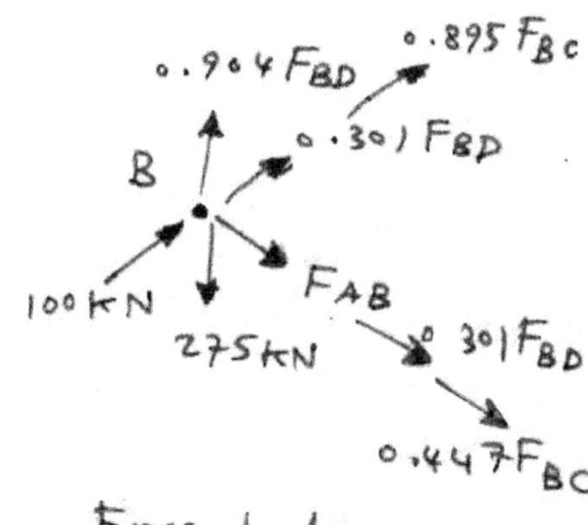

Substitute in eq. (2) :

$\quad 100 - 0.301(-359.5) - 0.895\,F_{AC} = 0$

$\quad\Rightarrow\ F_{AC} = \dfrac{232.6}{}\ kN\ (T).$

Substitute in eq. (1):

$\quad 0.301(-359.5) + 0.447(232.6) + F_{AB} = 0$

$\qquad\Rightarrow\ F_{AB} = 4.23\ kN\ (T).$

Joint B :

$\Sigma F_x = 0$:

$\quad \overset{4.23}{F_{AB}} + 0.301\,F_{BD} + 0.447\,F_{BC} = 0 \quad\text{——(1)}$

$\Sigma F_y = 0$:

$\quad 100 + 0.301\,F_{BD} + 0.895\,F_{BC} = 0 \quad\text{——(2)}$

$\Sigma F_z = 0$:

$\quad 0.904\,F_{BD} - 275 = 0$

$\qquad\Rightarrow\ F_{BD} = 304.2\ kN\ (T).$

Substitute in eq. (1) :

$\quad 4.23 + 0.301(304.2) + 0.447\,F_{BC} = 0$

$\qquad\Rightarrow\ F_{BC} = -214.3\ kN.$

$\qquad\quad \therefore\ F_{BC} = 214.3\ kN\ (C).$

Check with eq. (2):

$\quad 100 + 0.301(304.2) + 0.895(-214.3) = 0$

$\qquad\qquad -0.23 = 0 \ \checkmark$

Joint C :

$\sum F_x = 0$:

$$\underset{-214.3}{\underbrace{}}\quad \underset{232.6}{\underbrace{}}$$

$$200 + 0.447\, F_{BC} - 0.447\, F_{AC} = 0$$

$$\Rightarrow \quad 0.23 = 0 \quad \checkmark$$

$\sum F_y = 0$:

$$\underset{232.6}{\underbrace{}}\quad \underset{-214.3}{\underbrace{}}$$

$$0.316\, F_{CD} + 0.895\, F_{AC} + 0.895\, F_{BC} = 0$$

$$\Rightarrow \quad F_{CD} = -51.8 \text{ kN}.$$

$$\therefore \quad F_{CD} = 51.8 \text{ kN (C)}.$$

$\sum F_z = 0$ (Check) :

$$0.949\,(-51.8) + 50 = 0$$

$$0.8 = 0 \quad \checkmark$$

Final check at Joint D :

$\sum F_x = 0$:

$$0.301\, F_{BD} - 200 - 0.301\, F_{AD} = 0$$

$$0.301\,(304.2) - 200 - 0.301\,(-359.5) = 0$$

$$-0.2 = 0 \quad \checkmark$$

$\sum F_y = 0$:

$$0.301\, F_{BD} + 0.301\, F_{AD} - 0.316\, F_{CD} = 0$$

$$0.301\,(304.2) + 0.301\,(-359.5) - 0.316\,(-51.8) = 0$$

$$\Rightarrow \quad -0.3 = 0 \quad \checkmark$$

$\sum F_z = 0$:

$$-100 - 0.904\, F_{AD} - 0.904\, F_{BD} - 0.949\, F_{CD} = 0$$

$$-100 - 0.904\,(-359.5) - 0.904\,(304.2) - 0.949\,(-51.8) = 0$$

$$-0.85 = 0 \quad \checkmark$$

Example 4.2 :

Determine the member forces and the reactions for the truss shown in the figure.

Solution :

There are five joints.

We have three equations of equilibrium at each joint

$$\Sigma F_x = 0$$

$$\Sigma F_y = 0$$

$$\Sigma F_z = 0$$

∴ We have a total of fifteen equations.

Unknowns — reactions = 7

bar forces = 8

Total unknowns = 15

∴ The truss is determinate.

We will set up the fifteen equations in fifteen unknowns and use a computer program to solve them.

$$\boxed{3j = m + r} \quad \checkmark$$

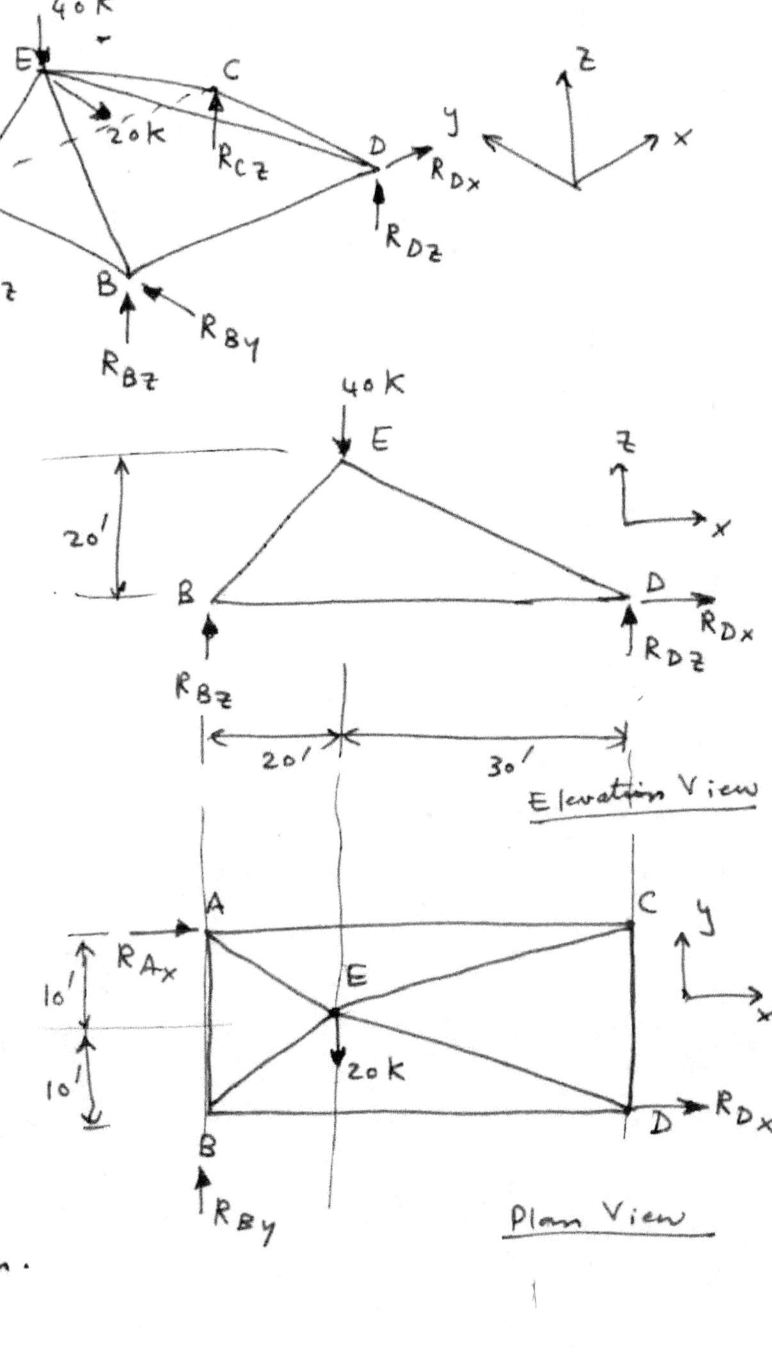

Member	Projection			Length	Ratio of projection to length (Direction Cosine)		
	x	y	z		x/L	y/L	z/L
AB	0	20	0	20	0	1	0
BD	50	0	0	50	1	0	0
DC	0	20	0	20	0	1	0
CA	50	0	0	50	1	0	0
AE	20	10	20	30	0.667	0.333	0.667
BE	20	10	20	30	0.667	0.333	0.667
DE	30	10	20	37.42	0.802	0.267	0.534
CE	30	10	20	37.42	0.802	0.267	0.534

Joint A :

$\Sigma F_x = 0 :$

$$R_{Ax} + F_{AC} + 0.667 F_{AE} = 0 \quad\text{———— (1)}$$

$\Sigma F_y = 0 :$

$$-F_{AB} - 0.333 F_{AE} = 0 \quad\text{———— (2)}$$

$\Sigma F_z = 0 :\; R_{Az} + 0.667 F_{AE} = 0$

$$\text{———— (3)}$$

Joint B :

$\Sigma F_x = 0 :$

$$F_{BD} + 0.667 F_{BE} = 0 \quad\text{———— (4)}$$

$\Sigma F_y = 0 :$

$$+R_{By} + F_{AB} + 0.333 F_{BE} = 0 \quad\text{———— (5)}$$

$\Sigma F_z = 0 :$

$$R_{Bz} + 0.667 F_{BE} = 0 \quad\text{———— (6)}$$

Joint C :

$\Sigma F_x = 0$:

$\quad -F_{AC} - 0.802\,F_{CE} = 0 \quad\text{——— (7)}$

$\Sigma F_y = 0$:

$\quad -F_{CD} - 0.267\,F_{CE} = 0 \quad\text{——— (8)}$

$\Sigma F_z = 0$:

$\quad R_{CZ} + 0.534\,F_{CE} = 0 \quad\text{——— (9)}$

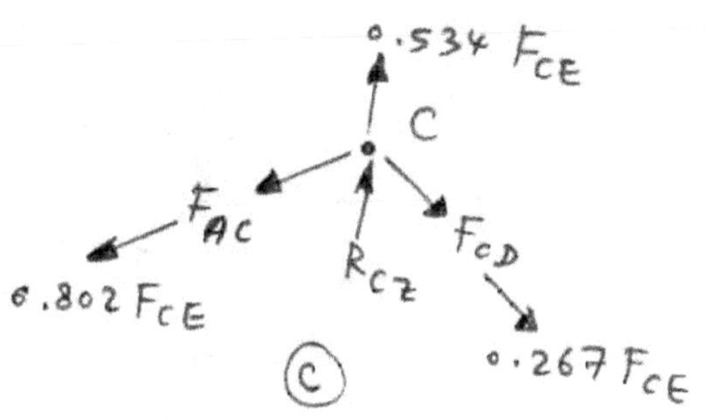

Joint D :

$\Sigma F_x = 0$:

$\quad R_{DX} - F_{BD} - 0.802\,F_{DE} = 0 \quad\text{——— (10)}$

$\Sigma F_y = 0$:

$\quad F_{CD} + 0.267\,F_{DE} = 0 \quad\text{——— (11)}$

$\Sigma F_z = 0$:

$\quad R_{DZ} + 0.534\,F_{DE} = 0 \quad\text{——— (12)}$

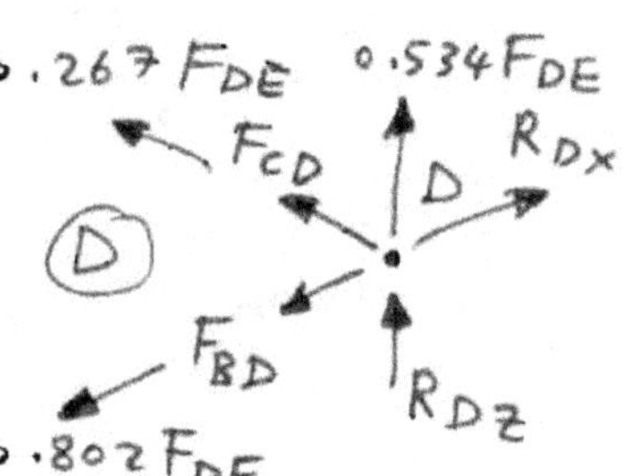

Joint E :

$\Sigma F_x = 0$:

$\quad 0.802\,F_{CE} + 0.802\,F_{DE} - 0.667\,F_{AE}$

$\qquad\quad - 0.667\,F_{BE} = 0 \quad\text{——— (13)}$

$\Sigma F_y = 0$:

$\quad 0.333\,F_{AE} - 20 - 0.333\,F_{BE} + 0.267\,F_{CE}$

$\qquad\quad - 0.267\,F_{DE} = 0 \quad\text{——— (14)}$

$\Sigma F_z = 0$:

$\quad -40 - 0.667\,F_{AE} - 0.667\,F_{BE} - 0.534\,F_{CE}$

$\qquad\quad - 0.534\,F_{DE} = 0 \quad\text{——— (15)}$

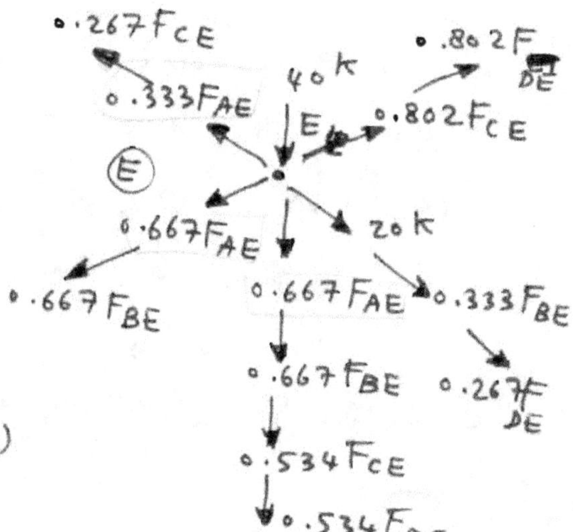

Write the equations in <u>matrix</u> form:

	F_{AB}	F_{BD}	F_{DC}	F_{AC}	F_{AE}	F_{BE}	F_{ED}	F_{CE}	R_{AZ}	R_{BZ}	R_{CZ}	R_{DZ}	R_{AX}	R_{BY}	R_{DX}
1	0	0	0	1	0.667	0	0	0	0	0	0	0	1	0	0
2	-1	0	0	0	-0.333	0	0	0	0	0	0	0	0	0	0
3	0	0	0	0	0.667	0	0	0	1	0	0	0	0	0	0
4	0	1	0	0	0	0.667	0	0	0	0	0	0	0	0	0
5	1	0	0	0	0	0.333	0	0	0	0	0	0	0	1	0
6	0	0	0	0	0	0.667	0	0	0	1	0	0	0	0	0
7	0	0	0	-1	0	0	0	-0.802	0	0	0	0	0	0	0
8	0	0	-1	0	0	0	0	-0.267	0	0	0	0	0	0	0
9	0	0	0	0	0	0	0	0.534	0	0	1	0	0	0	0
10	0	-1	0	0	0	0	0	-0.802	0	0	0	0	0	0	1
11	0	0	1	0	0	0	0	0.267	0	0	0	0	0	0	0
12	0	0	0	0	0	0	0	0.534	0	0	0	1	0	0	0
13	0	0	0	0	-0.667	-0.667	0.802	0.802	0	0	0	0	0	0	0
14	0	0	0	0	0.333	-0.333	-0.267	0.267	0	0	0	0	0	0	0
15	0	0	0	0	-0.667	-0.667	-0.534	0.534	0	0	0	0	0	0	0

Unknown vector:

$$\{F_{AB}, F_{BD}, F_{DC}, F_{AC}, F_{AE}, F_{BE}, F_{ED}, F_{CE}, R_{AZ}, R_{BZ}, R_{CZ}, R_{DZ}, R_{AX}, R_{BY}, R_{DX}\}$$

Right-hand side (=):

1	2	3	4	5	6	7	8	9	10	11	12	13	14	15
0	0	0	0	0	0	0	0	0	0	0	0	0	20	40

<u>input data</u> :

Enter the number of equations :
    Enter   15
Enter linear system one equation at a time :
Conclude each equation with its right-hand constant :
    Enter coefficients of equation 1

0,0,0,1,0.667,0,0,0,0,0,0,0,1,0,0,0
    Enter coefficients of equation 2

-1,0,0,0,-0.333,0,0,0,0,0,0,0,0,0,0,0
    Enter coefficients of equation 3

0,0,0,0,0.667,0,0,0,1,0,0,0,0,0,0,0
    Enter coefficients of equation 4

0,1,0,0,0,0.667,0,0,0,0,0,0,0,0,0,0
    Enter coefficients of equation 5

1,0,0,0,0,0.333,0,0,0,0,0,0,0,1,0,0

Enter coefficients of equation 6

0, 0, 0, 0, 0, 0.667, 0, 0, 0, 1, 0, 0, 0, 0, 0, 0

Enter coefficients of equation 7

0, 0, 0, -1, 0, 0, 0, -0.802, 0, 0, 0, 0, 0, 0, 0, 0

Enter coefficients of equation 8

0, 0, -1, 0, 0, 0, 0, -0.267, 0, 0, 0, 0, 0, 0, 0, 0

Enter coefficients of equation 9

0, 0, 0, 0, 0, 0, 0, 0.534, 0, 0, 1, 0, 0, 0, 0, 0

Enter coefficients of equation 10

0, -1, 0, 0, 0, 0, -0.802, 0, 0, 0, 0, 0, 0, 0, 1, 0

Enter coefficients of equation 11

0, 0, 1, 0, 0, 0, 0.267, 0, 0, 0, 0, 0, 0, 0, 0, 0

Enter coefficients of equation 12

0, 0, 0, 0, 0, 0, 0.534, 0, 0, 0, 0, 1, 0, 0, 0, 0

Enter coefficients of equation 13

0, 0, 0, 0, -0.667, -0.667, 0.802, 0.802, 0, 0, 0, 0, 0, 0, 0, 0

Enter coefficients of equation 14

0, 0, 0, 0, 0.333, -0.333, -0.267, 0.267, 0, 0, 0, 0, 0, 0, 0, 20

Enter coefficients of equation 15

0, 0, 0, 0, -0.667, -0.667, -0.534, -0.534, 0, 0, 0, 0, 0, 0, 0, 40

The <u>results</u> are:

$F_{AB} = -4\,k$	$R_{Az} = -8\,k$
$F_{BD} = 32\,k$	$R_{Bz} = 32\,k$
$F_{DC} = 4\,k$	$R_{Cz} = 8\,k$
$F_{AC} = 12\,k$	$R_{Dz} = 8\,k$
$F_{AE} = 12\,k$	$R_{Ax} = -20\,k$
$F_{BE} = -48\,k$	$R_{By} = 20\,k$
$F_{ED} = -15\,k$	$R_{Dx} = 20\,k$
$F_{CE} = -15\,k$	

Homework 3: 4.2, (4.4). Set up the equations and use the computer program to solve them.

Chapter 5: Shear and Moment Diagrams for Beams and Frames

5.1 Introduction:

* <u>Beams</u> are relatively long and slender members that are usually loaded normal to their longitudinal axis. They transfer loads by developing a combination of <u>bending</u> and <u>shear</u> stresses.

* <u>Types of Beams:</u>

simply-supported Beam
(determinate)

Cantilever Beam
(determinate)

Fixed-fixed Beam
(indeterminate to the <u>third</u> degree)

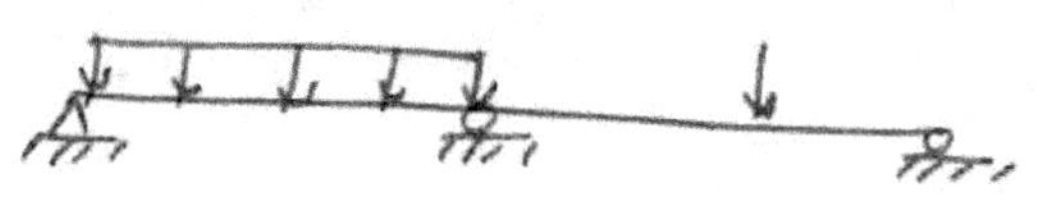

Continuous Beam
(indeterminate to the first degree)

5.2 Internal Forces:

* Internal forces are obtained by passing sections through the member and considering the free-body diagram on either side of the section.

Example:

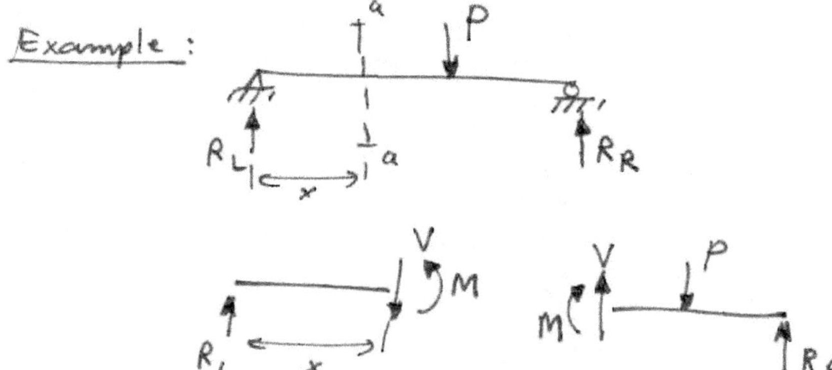

Free-body diagrams

V: Shear Force
M: Bending Moment

* the shear and bending moment diagrams are graphs of (36) the magnitude of the shear and bending moment plotted along the span of the member.

5.3 Sign Convention :

* A shear force is positive if it produces a clockwise moment about a point in the free-body.

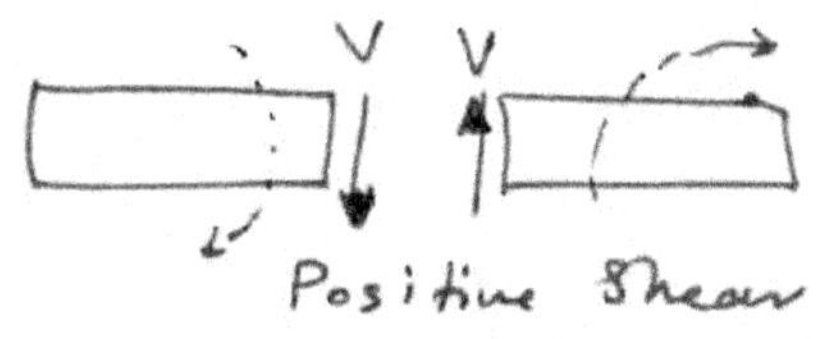

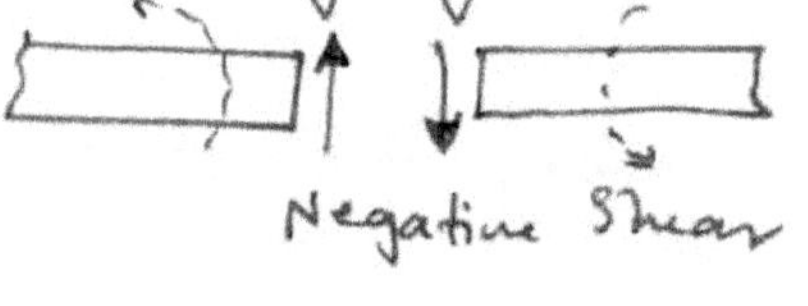

* the bending moment is positive if it causes compression in the upper fibers of the beam and tension in the lower fibers.

It causes the upper surface of the beam to take on a concave shape.

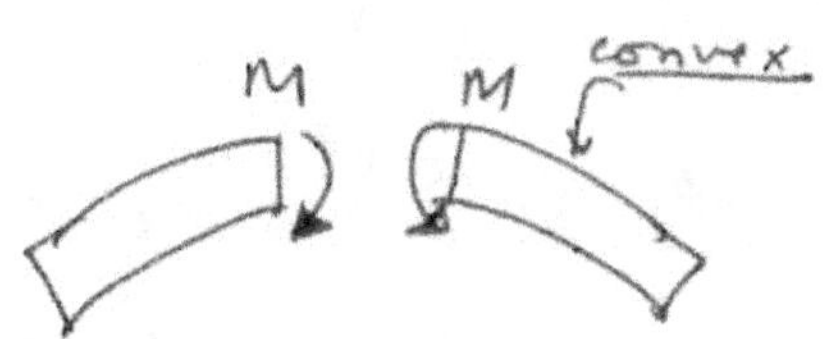

5.4 Shear and Bending—Moment Diagrams by the Method of Sections :

Example :

* Taking sections at regular intervals along the span of the beam.

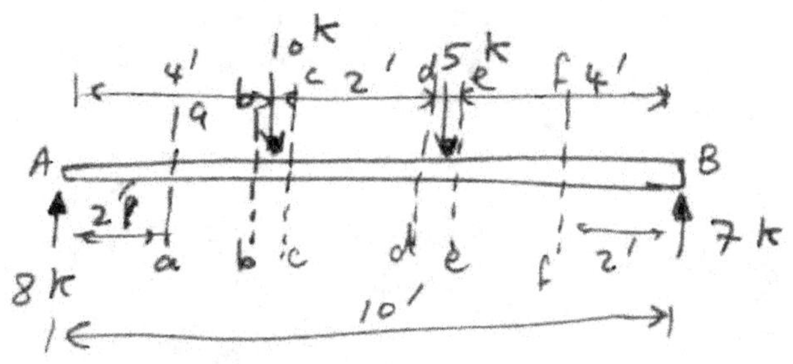

<u>Assume V and M to be positive.</u>

Section a–a:

$+\uparrow \ \Sigma F_y = 0 : \quad V = +8 \, k .$

$\Sigma M_A = 0 : \quad M - 8(2) = 0$

$$M = 16 \ k\text{-ft}.$$

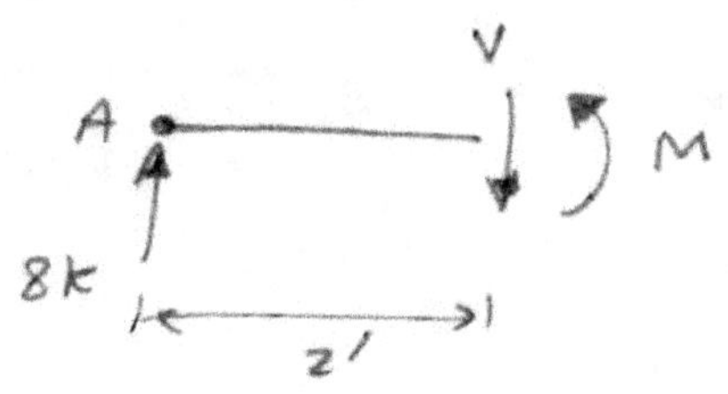

Section b–b:

$+\uparrow \ \Sigma F_y = 0 : \quad V = +8 k .$

$\Sigma M_A = 0 : \quad M - 8(4) = 0$

$$M = 32 \ k\text{-ft}.$$

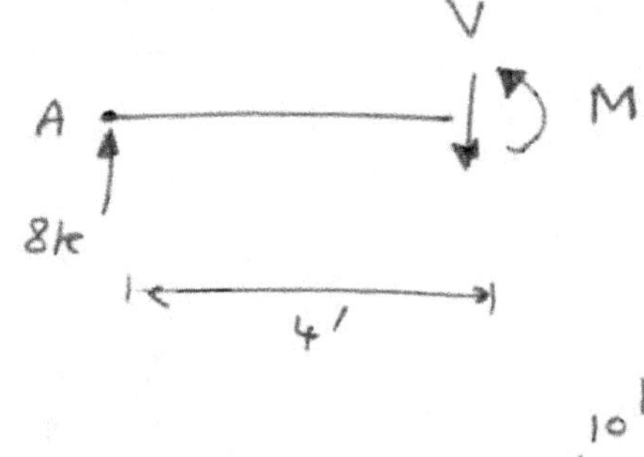

Section c–c:

$+\uparrow \ \Sigma F_y = 0 : \quad 8 - 10 - V = 0$

$$\Rightarrow V = -2 \ k .$$

$\Sigma M_A = 0 : \quad M - 10(4) - (-2)(4) = 0$

$$M = +32 \ k\text{-ft}.$$

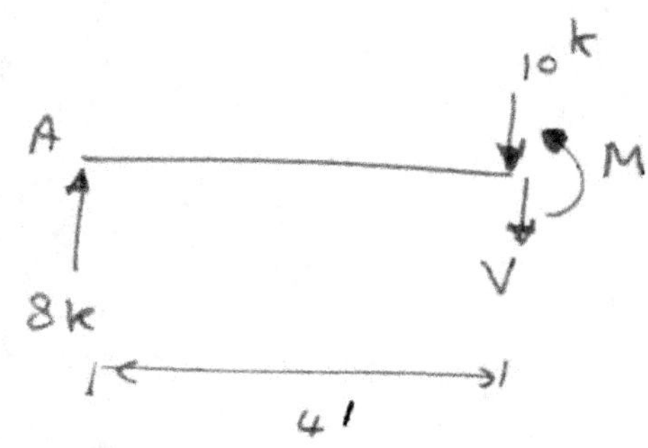

Section d–d:

$+\uparrow \ \Sigma F_y = 0 : \quad V + 7 - 5 = 0$

$$\therefore V = -2k .$$

$\Sigma M_B = 0 : \quad -M - (-2)(4) + 5(4) = 0$

$$\therefore M = 28 \ k\text{-ft}.$$

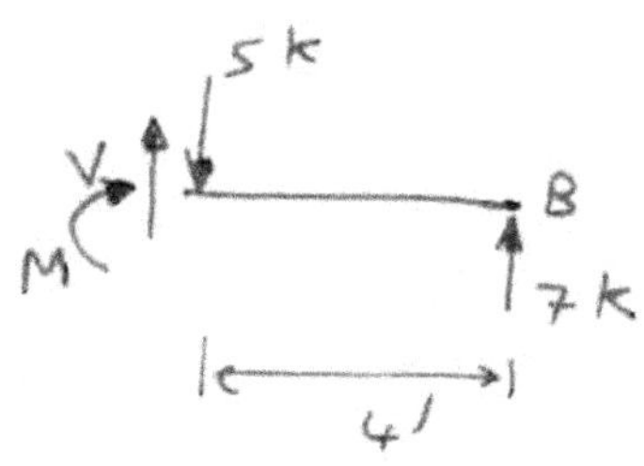

Section e–e:

$+\uparrow \ \Sigma F_y = 0 :$

$\quad V + 7 = 0, \quad \therefore V = -7 k .$

$\Sigma M_B = 0 : \quad -M - (-7)(4) = 0$

$$\therefore M = 28 k .$$

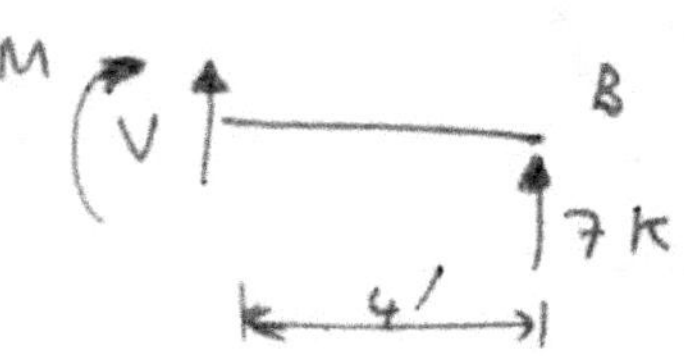

Section f–f:

$+\uparrow \ \Sigma F_y = 0 : \quad V = -7 k .$

$\Sigma M_B = 0 : \quad -M - (-7)(2) = 0 \Rightarrow M = 14 K .$

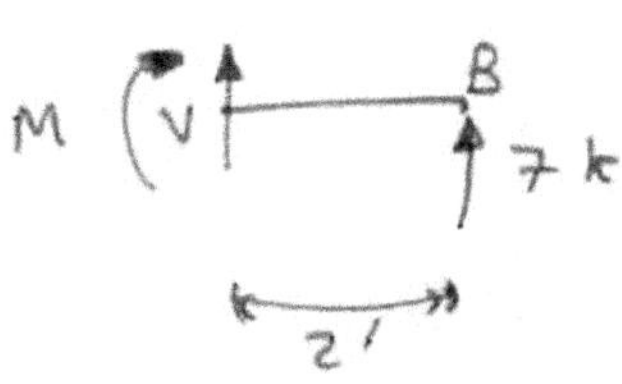

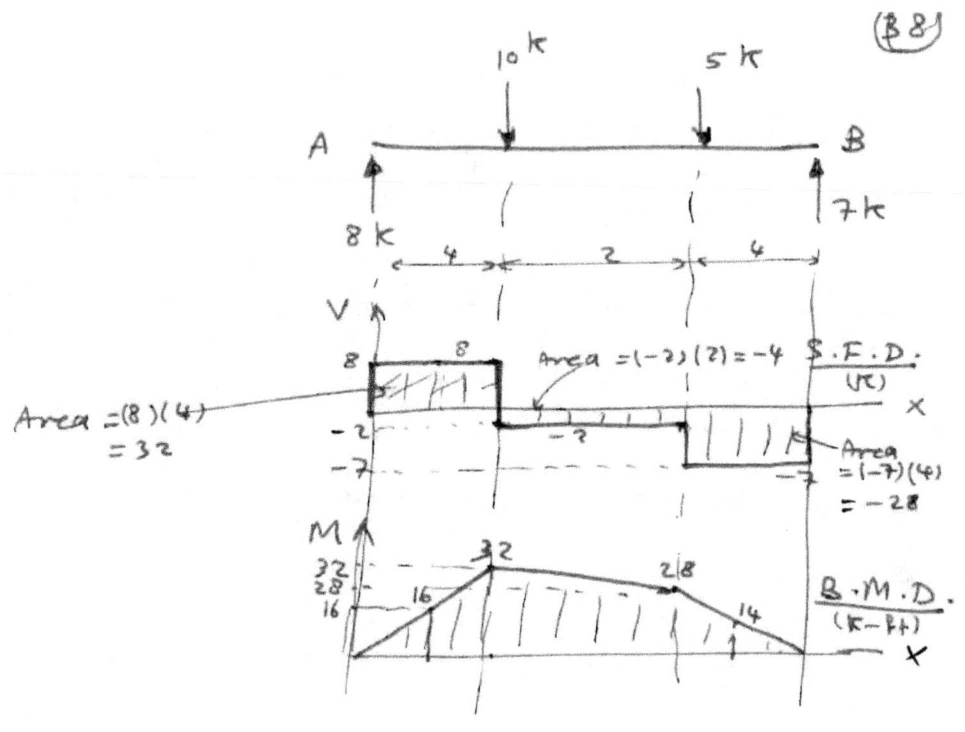

5.5 Relationships between Load, Shear, and Bending Moment:

<u>Example</u>:

Consider the differential element shown:

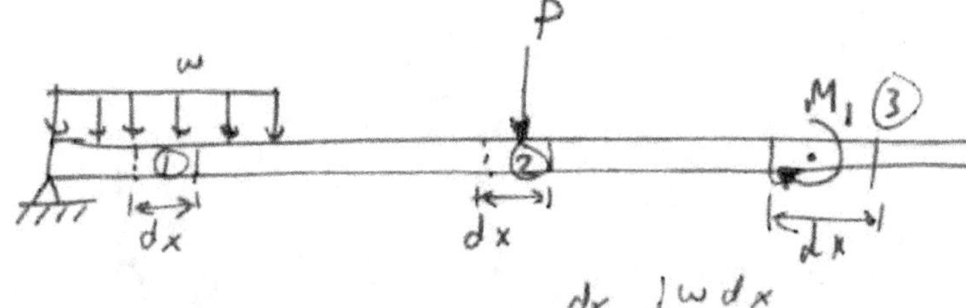

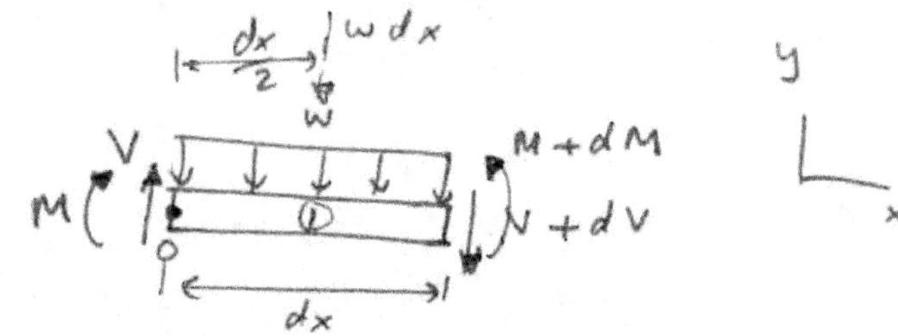

$+\uparrow \Sigma F_y = 0$:

$$V - (V + dV) - w\,dx = 0$$

$$V - V - dV - w\,dx = 0$$

$$dV = -w\,dx$$

$$\boxed{\dfrac{dV}{dx} = -w}$$

The change in shear between two points along the member is equal to the load between the points.

Minus sign ≡ decrease in shear.

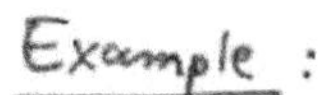

$\stackrel{+}{\curvearrowright} \Sigma M_0 = 0$:

$$-w\,dx \cdot \left(\dfrac{dx}{2}\right) - (V + dV)\,dx - M + (M + dM) = 0$$

$$-\dfrac{w}{2}(dx)^2 - V(dx) - \underbrace{(dV)(dx)}_{\text{neglected}} + dM = 0$$

$\underbrace{\phantom{-\dfrac{w}{2}(dx)^2}}_{\text{neglected}}$

$$-V\,dx = -dM$$

$$\therefore \boxed{\dfrac{dM}{dx} = V}$$ the slope of the moment diagram at any point is equal to the shear at that point.

The change in moment between two points is equal to the area of the shear diagram between the points.

Consider the differential element:

$+\uparrow \ \Sigma F_y = 0:$

$$\cancel{V} - P - (\cancel{V} + dV) = 0$$

$$\boxed{dV = -P}$$

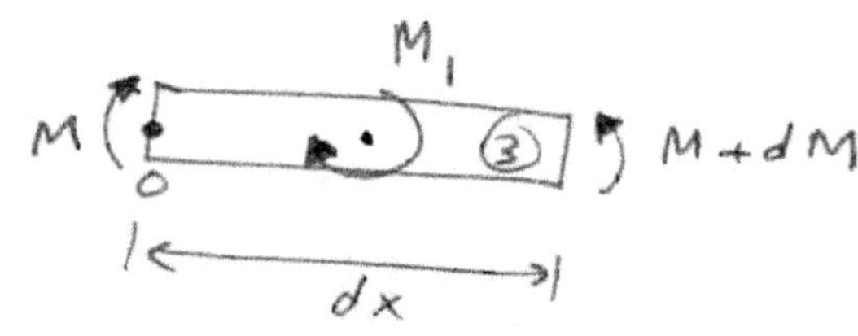

the shear decreases by an amount equal to the concentrated load. (a concentrated load results in an abrupt change in the shear force diagram).

Consider the differential element:

$+\curvearrowright \ \Sigma M_0 = 0:$

$$-\cancel{M} + (\cancel{M} + dM) - M_1 = 0 :$$

$$\boxed{dM = M_1}$$

A couple results in an abrupt change of the bending moment diagram.

5·6 Construction of Shear and Moment Diagrams:

Example 5·1 :

Construct the shear and bending-moment diagram for the beam shown in the figure.

Solution:

(1) Calculate the reactions:

$\xrightarrow{+} \Sigma F_x = 0: \ R_2 = 0 .$

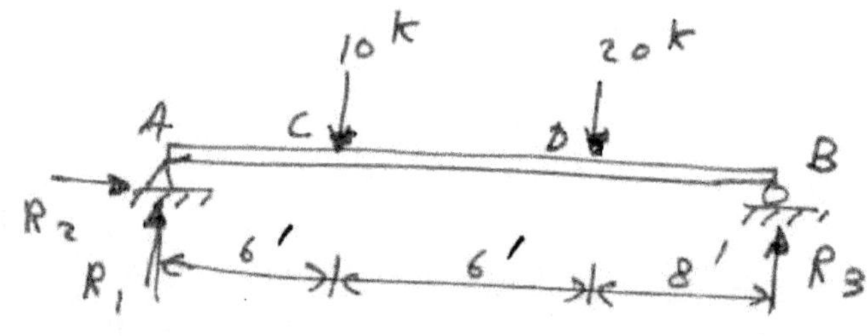

$\curvearrowright \; \Sigma M_A = 0 : \quad -10(6) - 20(12) + R_3(20) = 0$

$$\therefore \quad R_3 = 15^k \uparrow .$$

$+\uparrow \; \Sigma F_y = 0 : \quad R_1 + 15 - 10 - 20 = 0$

$$\therefore \quad R_1 = 15^k .$$

Check the results:

$\curvearrowright \; \Sigma M_B = 0 : \quad -15(20) + 10(14) + (20)(8) = 0$

$$0 = 0 .$$

Show all the reactions on a diagram.

<u>Section a–a</u> :

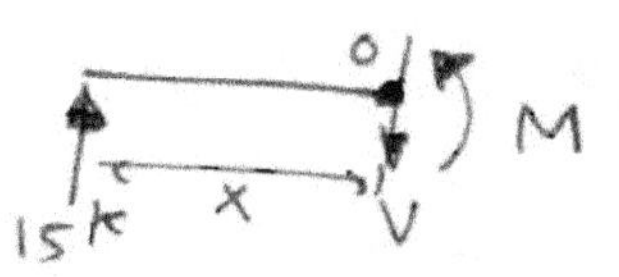

$\curvearrowright \; \Sigma M_0 = 0 : \quad M - 15x = 0$

$$\therefore \quad M(x) = 15x , \qquad 0 \le x \le 6 .$$

$$M(6) = 15(6) = 90 \text{ k-ft} .$$

Area of Shear force diagram $= 15(6) = 90$ k-ft.

Change in moment between C and D is equal to the area of the shear force diagram between C and D.

$$\text{Area} = +5(6) = 30 .$$

$$\therefore \quad M(12) = 90 + 30 = 120 \text{ k-ft} .$$

Between D and B, Area of S.F.D. $= -15(8) = -120 .$

$$\therefore \quad M(20) = 120 - 120 = 0 .$$

The shear is constant, the slope of the bending moment diagram is constant

___

<u>Note</u> : The maximum moment occurs where the shear force changes sign.

Example 5.2 :

Construct the shear and moment diagrams for the beam in the figure.

Solution :

(1) Find the reactions :

$$\xrightarrow{} \Sigma F_x = 0 : \quad R_2 = 0 .$$

$$\stackrel{+}{\curvearrowright} \Sigma M_A = 0 :$$
$$R_3(8) - 240(2) = 0$$
$$\therefore \quad R_3 = 60 \ kN \uparrow$$

$$+\uparrow \Sigma F_y = 0 : \quad R_1 + 60 - 240 = 0$$
$$\therefore \quad R_1 = 180 \ kN \uparrow .$$

Show the reactions on a diagram.

The change in shear between two points is equal to the area of the load diagram between the two points.

Area of load between A and B
$$= -20(4) = -80 \ kN .$$

Area of load between B and C
$$= -20(8) = -160 .$$
$$\therefore \quad V = +100 - 160 = -60 \ kN .$$

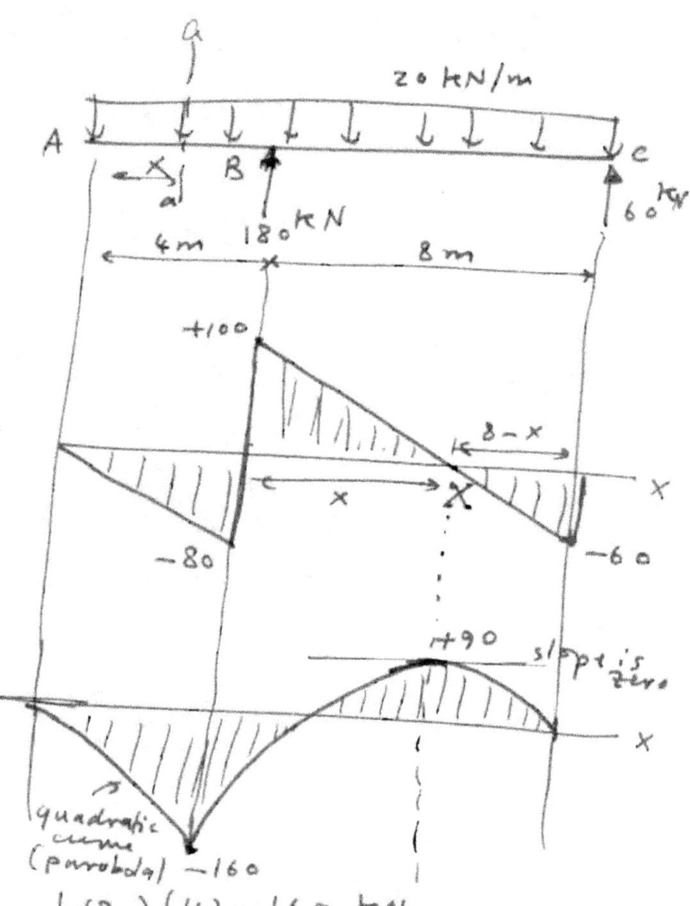

Area of S.F.D. between A and B $= -\frac{1}{2}(80)(4) = -160 \ kN\text{-}m .$

$$\stackrel{+}{\curvearrowright} \Sigma M_0 = 0 :$$
$$M + 20x \left(\frac{x}{2}\right) = 0$$
$$\therefore \quad M(x) = -10x^2 , \quad 0 \le x \le 4 \ m$$
$$M(4) = -10(4)^2 = -160 \ kN\text{-}m .$$

Calculate x : From similar triangles : $\dfrac{x}{8-x} = \dfrac{100}{60} \implies 60x = 800 - 100x$
$160x = 800 \implies x = 5 \ m .$

Area of S.F.D. between B and X
$$= +\frac{1}{2}(100)(5) = 250 \ kN\cdot m .$$

$$\therefore M_X = -160 + 250 = 90 \ kN \cdot m.$$

Area of S.F.D. between X and C = $-\frac{1}{2}(60)(3) = -90 \ kN \cdot m.$

$$M_C = +90 - 90 = 0.$$

Example 5.3 :

Draw the shear and moment diagram for the beam shown in the figure.

Solution :

(1) Calculate the reactions:

$\xrightarrow{+} \Sigma F_X = 0 : \quad R_2 = 0.$

$\curvearrowright \ \Sigma M_A = 0 :$

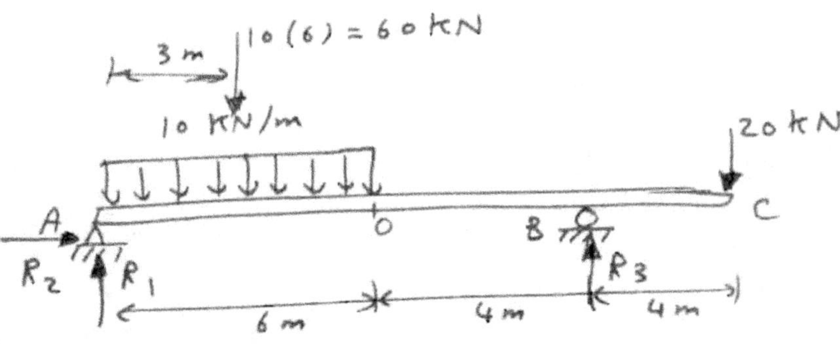

$$-60(3) + R_3(10) - 20(14) = 0$$

$$\therefore \quad R_3 = 46 \ kN \ \uparrow$$

$+\uparrow \ \Sigma F_y = 0 :$

$$R_1 + 46 - 60 - 20 = 0$$

$$\therefore \quad R_1 = 34 \ kN \ \uparrow.$$

S.F.D. :

$$V_O = 34 - 10(6) = -26 \ kN.$$

$$V_B = -26 + 46 = 20 \ kN.$$

Distance to zero shear :

From similar triangles:

$$\frac{34}{x} = \frac{26}{6-x}$$

$$26 x = 34(6) - 34x$$

$$x = 3.4 \ m.$$

B.M.D.

M_{max} = Area of triangle

$$= \frac{1}{2}(3.4)(34) = 57.8 \ kN \cdot m$$

$$M_O = 57.8 - \frac{1}{2}(6 - 3.4)(26)$$

$$= 24 \ kN \cdot m.$$

$$M_B = 24 - (26)(4) = -80 \ kN \cdot m.$$

$$M_C = -80 + (20)(4) = 0.$$

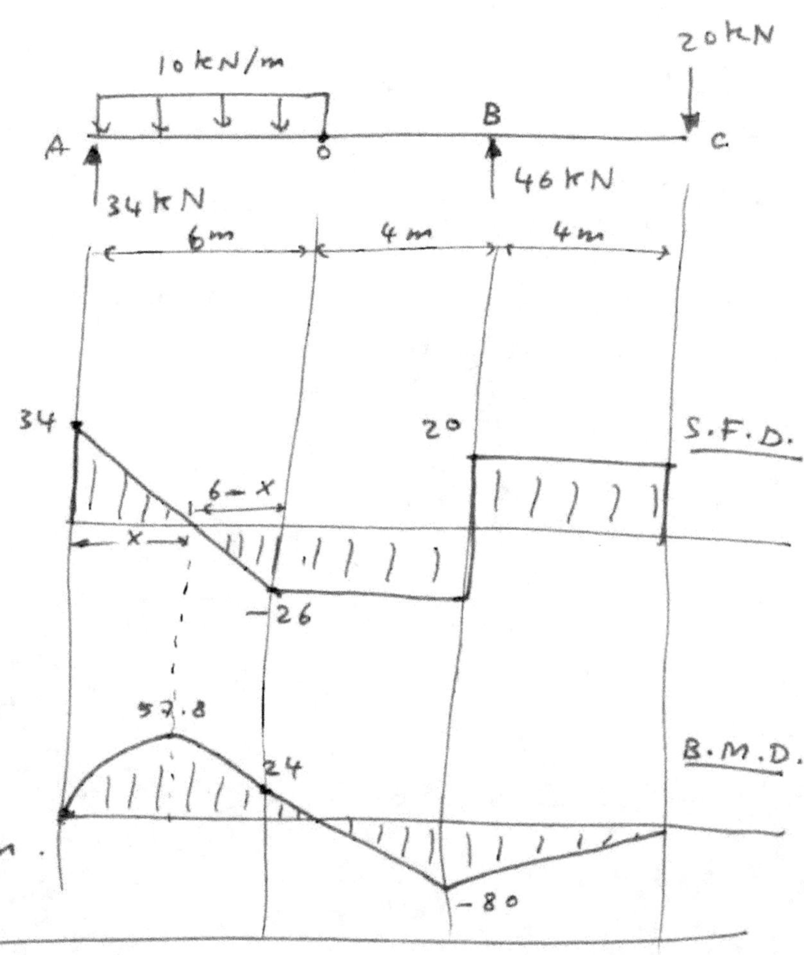

Example 5.4 :

Construct the shear and moment diagram for the beam shown in the figure.

Solution :

(1) Calculate the reactions :

$\xrightarrow{+} \Sigma F_x = 0 : \quad R_2 = 0 .$

$\stackrel{+}{\curvearrowleft} \Sigma M_B = 0 :$

$\qquad R_3(12) - 36(16-12) = 0$

$\qquad \therefore R_3 = 12\,k \uparrow .$

$+\uparrow \Sigma F_y = 0 :$

$\qquad R_1 + 12 - 36 = 0$

$\qquad \therefore R_1 = 24\,k \uparrow .$

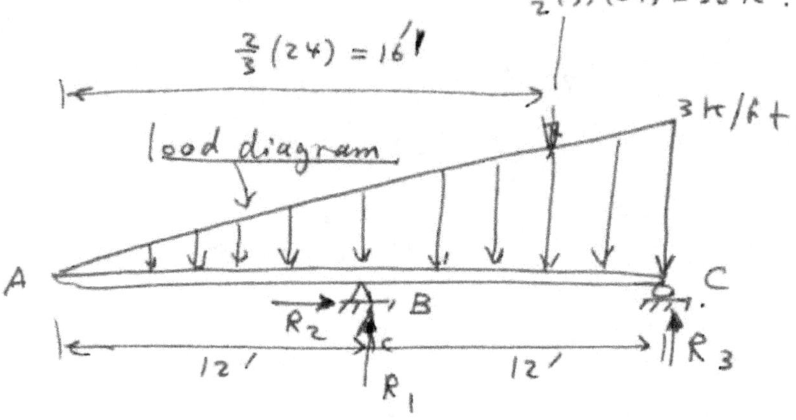

S.F.D. :

V_B = Area of load diagram

$\dfrac{W_B}{12} = \dfrac{3}{24} \Rightarrow W_B = 1.5\,k/ft .$

$\therefore V_{B_{left}} = -\frac{1}{2}(12)(1.5) = -9\,kN .$

$V_{B_{right}} = -9 + 24 = 15\,k .$

$V_C = 15 - \left(\dfrac{1.5+3}{2}\right)(12) = -12\,k .$

Distance to zero shear :

Take a section at point O :

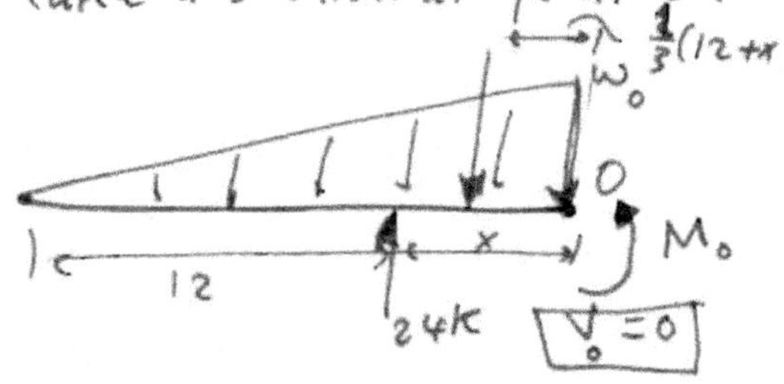

$\dfrac{W_0}{12+x} = \dfrac{3}{24}$

$\Rightarrow W_0 = \dfrac{12+x}{8}\,k/ft .$

Total vertical load

$= \frac{1}{2}\left(\dfrac{12+x}{8}\right)(12+x)$

$= \dfrac{(12+x)^2}{16} .$

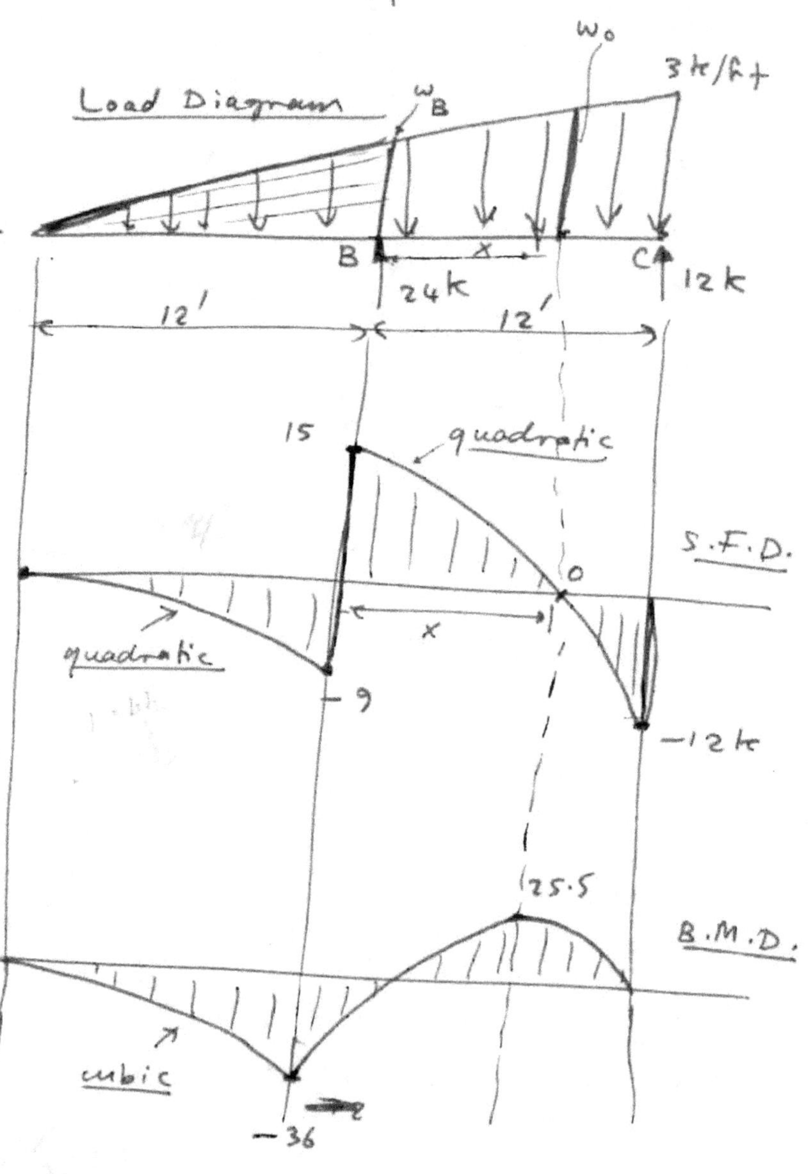

$+\uparrow \; V_0 = 0:$

$$24 - \frac{(12+x)^2}{16} = 0 \implies (12+x)^2 = 384$$

$$\therefore \quad 12 + x = 19.6$$

$$\therefore \quad x = 7.6'.$$

B.M.D.

$+\curvearrowright \Sigma M_0 = 0 \implies M_0 + \frac{(12+x)^2}{16} \cdot \frac{1}{3}(12+x) - 24x = 0$

$$\text{substitute} \quad x = 7.6' \quad , \quad 12+x = 19.6'.$$

$$\implies M_0 = 25.5 \; k\text{-}ft.$$

Section at B:

$+\curvearrowright \Sigma M_B = 0:$

$$M_B + 9\left(\frac{1}{3}\right)(12) = 0$$

$$\therefore \quad M_B = -36 \; k\text{-}ft.$$

check:

$+\curvearrowright \Sigma M_C = 0:$

$$\frac{1}{2}(3)(24)\left(\frac{1}{3}\right)(24) - 24(12) = 0$$

$$0 = 0 \checkmark$$

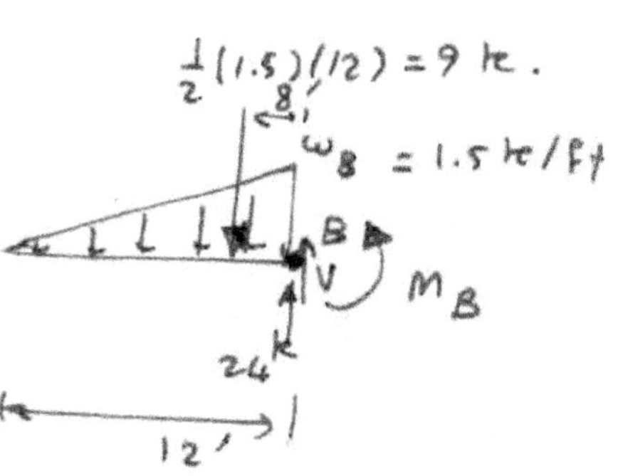

Example 5.5:

Construct the shear and moment diagrams for member ABCD of the structure in the figure.

Solution:

(1) Calculate the reactions.

Separate the structure into two parts at the hinge.

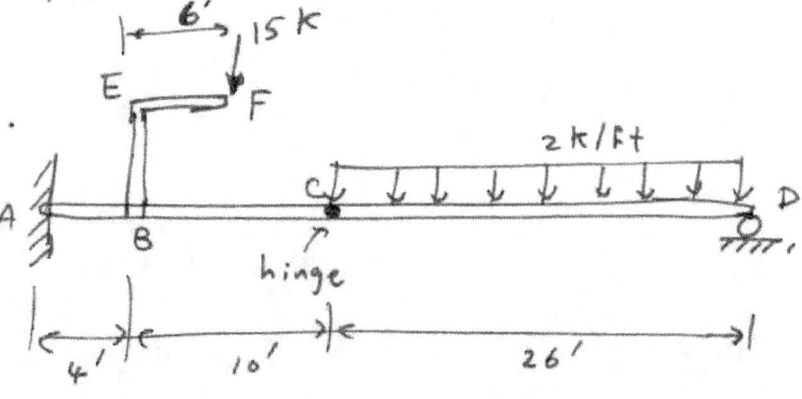

Member CD:

$\xrightarrow{+}\ \Sigma F_x = 0:\quad H_c = 0.$

$+\circlearrowleft\ \Sigma M_c = 0:$

$\qquad -52(13) + R_D(26) = 0$

$\qquad \therefore\ R_D = 26\ k\ \uparrow$

$+\uparrow\ \Sigma F_y = 0:\quad V_c + 26 - 52 = 0$

$\qquad \therefore\ V_c = 26\ k.$

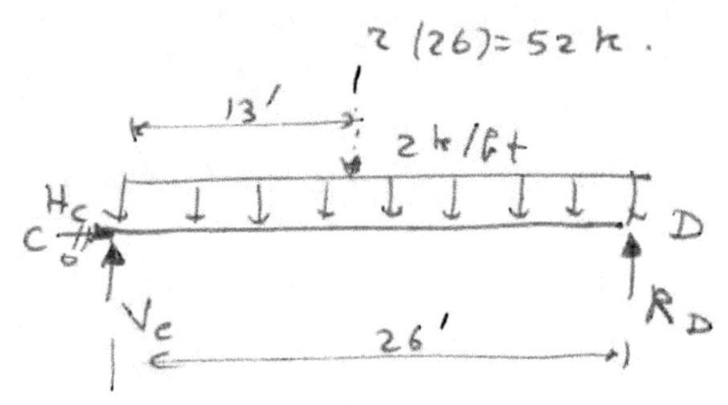

Member ABC:

$\xrightarrow{+}\ \Sigma F_x = 0:\quad R_{Ax} = 0.$

$+\uparrow\ \Sigma F_y = 0:$

$\qquad R_{Ay} - 26 - 15 = 0$

$\qquad \therefore\ R_{Ay} = 41\ k\ \uparrow.$

$+\circlearrowleft\ \Sigma M_A = 0:$

$\qquad M_A - 26(14) - 15(10) = 0$

$\qquad \therefore\ M_A = 514\ k\text{-}ft.$

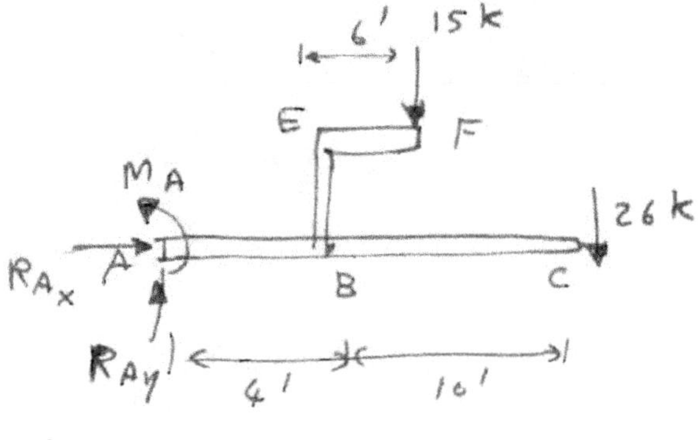

(2) Determine the internal forces at B:

Member BEF:

$\xrightarrow{+}\ \Sigma F_x = 0:\quad N_B = 0.$

$+\uparrow\ \Sigma F_y = 0:\quad V_B = 15\ k.$

$+\circlearrowleft\ \Sigma M_B = 0:\quad M_B - 15(6) = 0$

$\qquad \therefore\ M_B = 90\ k\text{-}ft.$

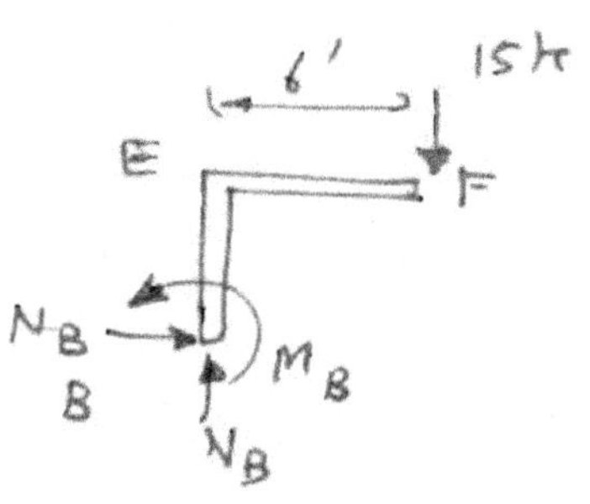

Put the reactions at B as applied loads on the structure.

$\underline{S.F.D.:}$

$V_A = +41 \text{ k.}$

$V_B = 41 - 15 = 26 \text{ k.}$

$V_D = 26 - (2 \times 26)$
$\quad = -26 \text{ k.}$

$\underline{B.M.D.:}$

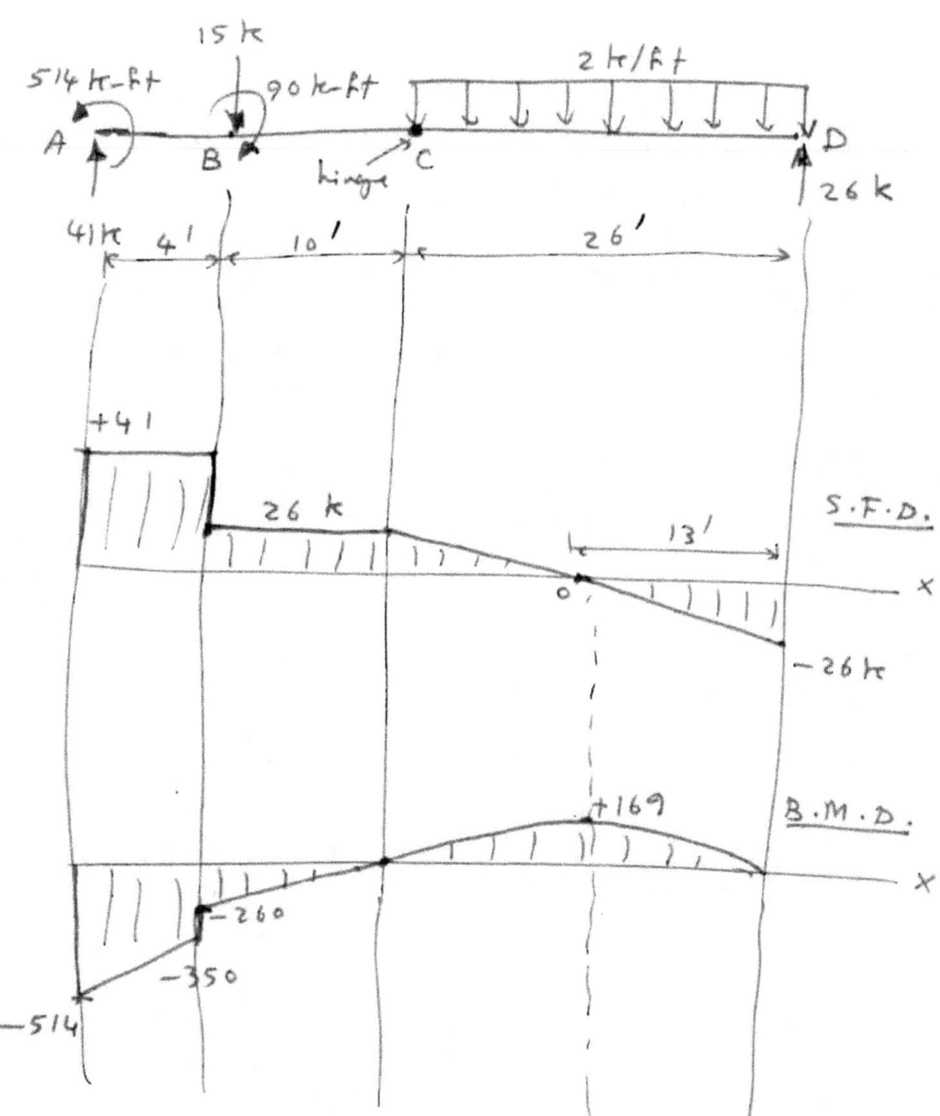

514 k-ft

$A \qquad M_A$

$M_A = -514 \text{ k-ft.}$

$\bar{M_B} = -514 + (41)(4)$
$\quad = -350 \text{ k-ft.}$

90 k-ft

$350 \text{ k-ft} \qquad M_B^+$

$M_B^+ + 350 - 90 = 0$

$\therefore M_B^+ = -260 \text{ k-ft.}$

$M_C = -260 + (26)(10)$
$\quad = 0.$

$M_{o \atop (max)} = 0 + \frac{1}{2}(13)(26)$
$\quad = +169 \text{ k-ft.}$

$M_D = 169 - \frac{1}{2}(13)(26)$
$\quad = 0.$

* A __rigid frame__ consists of flexural members connected to one another in such a manner that the members __cannot__ __rotate__ in relation to each other at the joints.

* Moments and forces can be transferred at the joints.

Example 5.6 :

Construct the __normal__ $\overset{(axial)}{}$ force, __shear__ force and __bending moment__ diagram for each member of the frame shown in the figure.

Solution:

(1) Calculate the reactions:

$\overset{+}{\to} \Sigma F_x = 0$:

$-R_1 + 60 + 40 + 40 = 0$

$R_1 = 140 \ kN \leftarrow$

$\curvearrowright \Sigma M_A = 0$:

$-60(2) - 40(4) - 40(4)$

$- 280(7) + R_3(14) = 0$

$\Rightarrow R_3 = 171.4 \ kN \uparrow.$

$+\uparrow \Sigma F_y = 0$:

$R_2 + 171.4 - 280 = 0$

$\therefore R_2 = 108.6 \ kN \uparrow.$

(2) Consider each member of the frame separately:

Member __AB__ :

$\overset{+}{\to} \Sigma F_x = 0$:

$-140 + 60 + B_1 = 0$

$\therefore B_1 = 80 \ kN \to$

$+\uparrow \Sigma F_y = 0$:

$-B_2 + 108.6 = 0$

$B_2 = 108.6 \ kN \downarrow$

$\circlearrowleft \; \Sigma M_B = 0:$

$\qquad M_B + 60(2) - 140(4) = 0$

$\qquad \therefore \; M_B = 440 \; kN \cdot m \; \circlearrowright$

<u>Joint B</u> : (Consider a joint only when there is an external applied load or moment acting directly on the joint).

$\qquad M_B = 440 \; kN \cdot m \; \circlearrowright$

$\xrightarrow{+} \; \Sigma F_x = 0:$

$\qquad -80 + 40 + X_1 = 0$

$\qquad \therefore \; X_1 = 40 \; kN \; \rightarrow$

$+\uparrow \; \Sigma F_y = 0:$

$\qquad 108.6 - X_2 = 0 \; \Rightarrow X_2 = 108.6 \; kN \downarrow .$

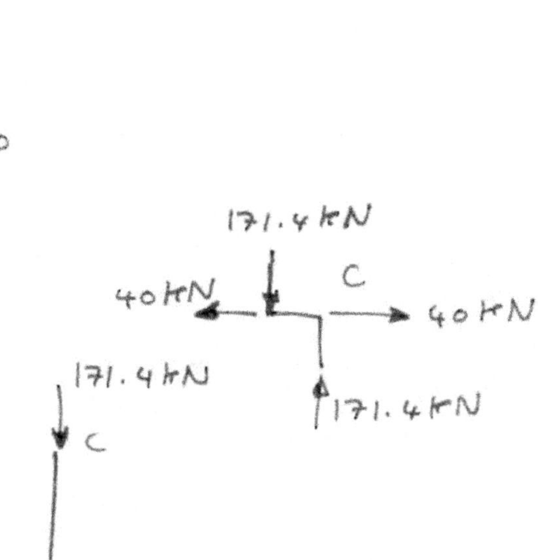

<u>Member BC</u> :

$\xrightarrow{+} \; \Sigma F_x = 0:$

$\qquad -40 + N_c = 0:$

$\qquad N_c = 40 \; kN \; \rightarrow$

$+\uparrow \; \Sigma F_y = 0:$

$\qquad 108.6 - 280 + V_c = 0$

$\qquad V_c = 171.4 \; kN \uparrow .$

$\circlearrowleft \; \Sigma M_c = 0:$

$\qquad M_c - 440 + 280(7) - 108.6(14) = 0$

$\qquad \therefore \; M_c = 0.4 \approx 0 .$

<u>Joint C</u> : (there is an applied load) (no moment) .

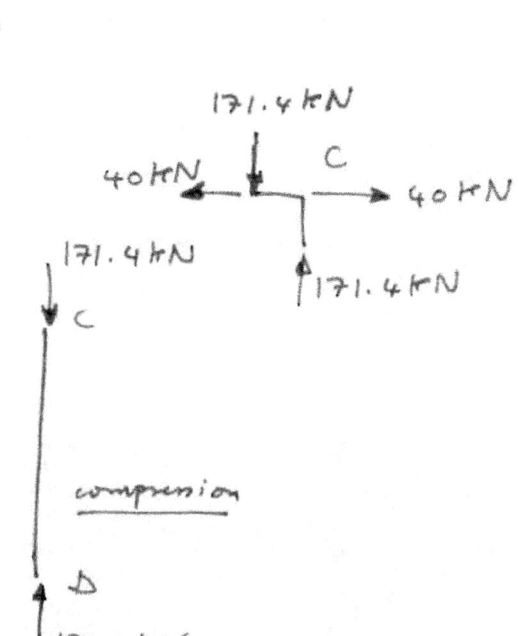

<u>Member CD</u> :

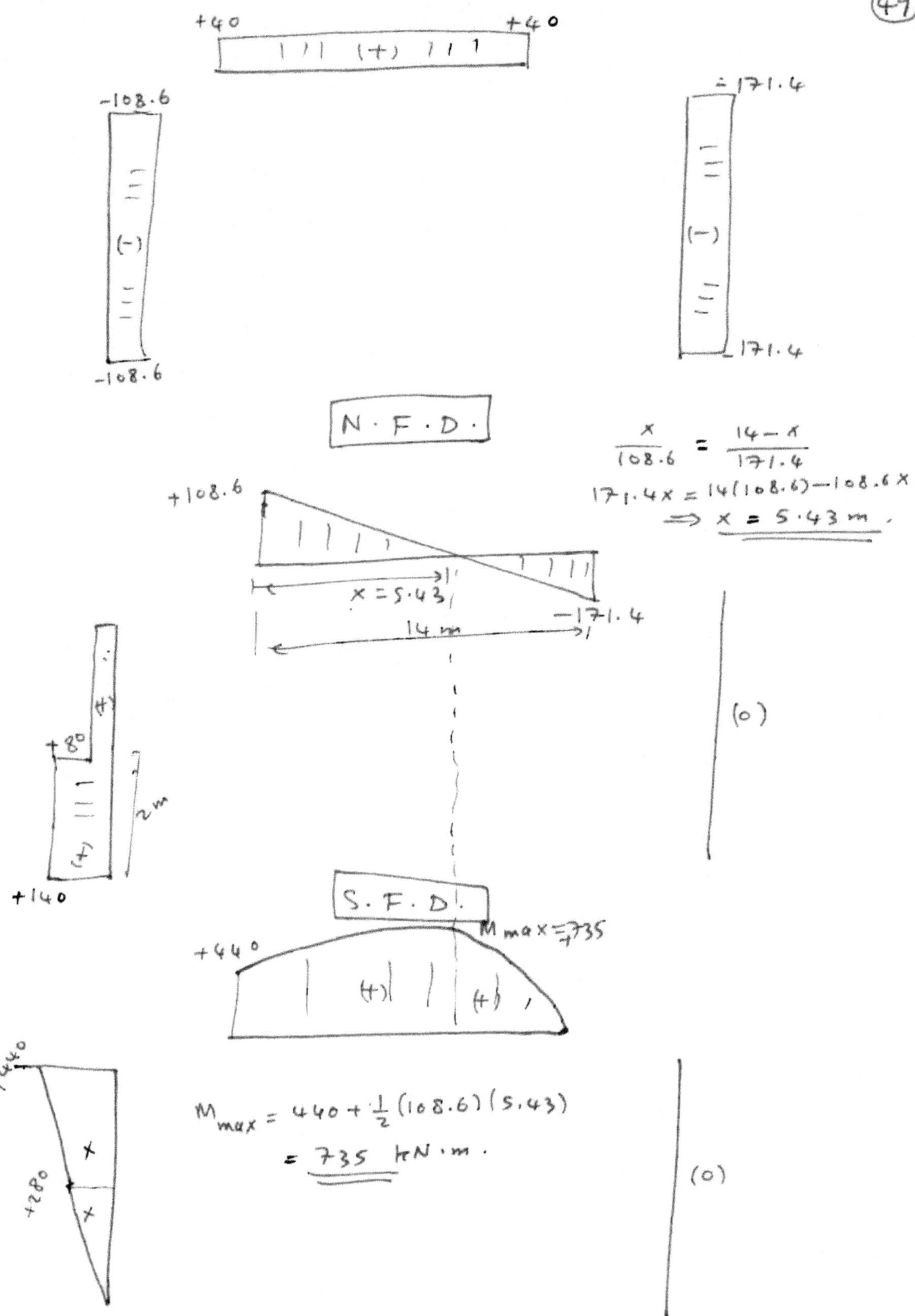

$M_{max} = 440 + \dfrac{1}{2}(108.6)(5.43)$

$= 735\ kN\cdot m$

$M_1 = 140(2) = 280\ kN\cdot m$

* the moment diagram is drawn on the compression side of the member.

Example 5.7 :

Construct the axial (normal), shear and moment diagrams for each member of the frame shown in the figure.

Solution :

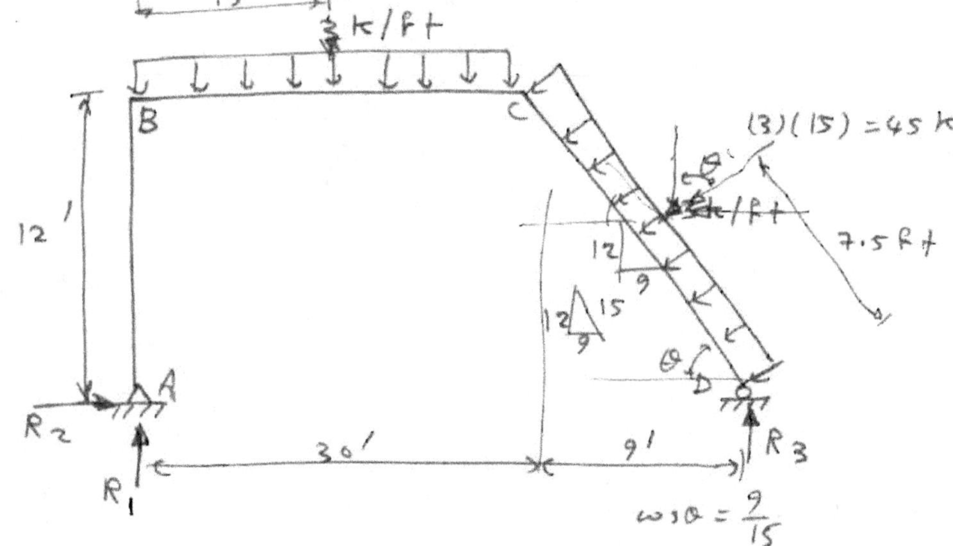

(1) Calculate the reactions ;

$\xrightarrow{+} \Sigma F_x = 0 :$
$$R_2 - (45)\left(\frac{12}{15}\right) = 0 \implies R_2 = 36 \text{ k} \rightarrow$$

$\circlearrowleft^+ \Sigma M_D = 0 :$
$$45(7.5) + 90(9+15) - R_1(9+30) = 0$$
$$\therefore R_1 = 64.04 \text{ k} \uparrow$$

$+\uparrow \Sigma F_y = 0 : \quad 64.04 + R_3 - 90 - (45)\left(\frac{9}{15}\right) = 0$
$$\therefore R_3 = 52.96 \text{ k} \uparrow$$

(2) Consider each member of the frame separately:

Member AB :

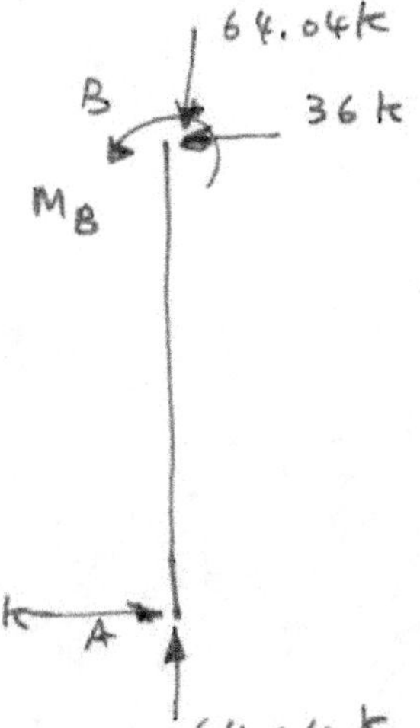

$\circlearrowleft^+ \Sigma M_B = 0 :$
$$M_B + 36(12) = 0$$
$$\therefore M_B = -432 \text{ k-ft}$$

Member BC :

$+\uparrow \Sigma F_y = 0:$

$V_c + 64.04 - 90 = 0$

$V_c = 25.96 \text{ k} \uparrow .$

$+\circlearrowleft \Sigma M_c = 0:$

$M_c + 432 - 64.04(30) + 90(15) = 0$

$\therefore M_c = +139.2 \text{ k-ft} .$

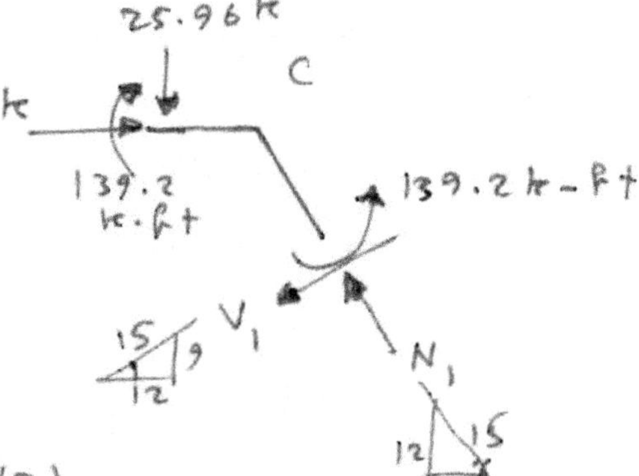

Joint C :

$\rightarrow \Sigma F_x = 0:$

$36 - V_1\left(\frac{12}{15}\right) - N_1\left(\frac{9}{15}\right) = 0 \quad\text{—— (1)}$

$+\uparrow \Sigma F_y = 0:$

$-25.96 - V_1\left(\frac{9}{15}\right) + N_1\left(\frac{12}{15}\right) = 0 \quad\text{——(2)}$

Solve the two equations simultaneously:

From eq.(1):

$V_1 = \frac{15}{12}\left(36 - \frac{9}{15} N_1\right)$

substitute in eq.(2):

$-25.96 - \left(\frac{9}{15}\right)\left(\frac{15}{12}\right)\left(36 - \frac{9}{15} N_1\right) + \frac{12}{15} N_1 = 0$

$-52.96 + 1.25 N_1 = 0$

$\therefore N_1 = 42.37 \text{ K} .$

$\therefore V_1 = \frac{15}{12}\left(36 - \frac{9}{15}(42.37)\right) = 13.22 \text{ k} .$

Member CD :

$+\nearrow \Sigma F_y = 0:$

$13.22 + V_D - 45 = 0$

$\therefore V_D = 31.78 \text{ k} .$

$+\circlearrowleft \Sigma M_D = 0:$

$-139.2 - 13.22(15) + 45(7.5) = 0$

$0 = 0 \checkmark$

Check: $\sqrt{(42.37)^2 + (31.78)^2} = 52.96 \text{ k} . \checkmark$

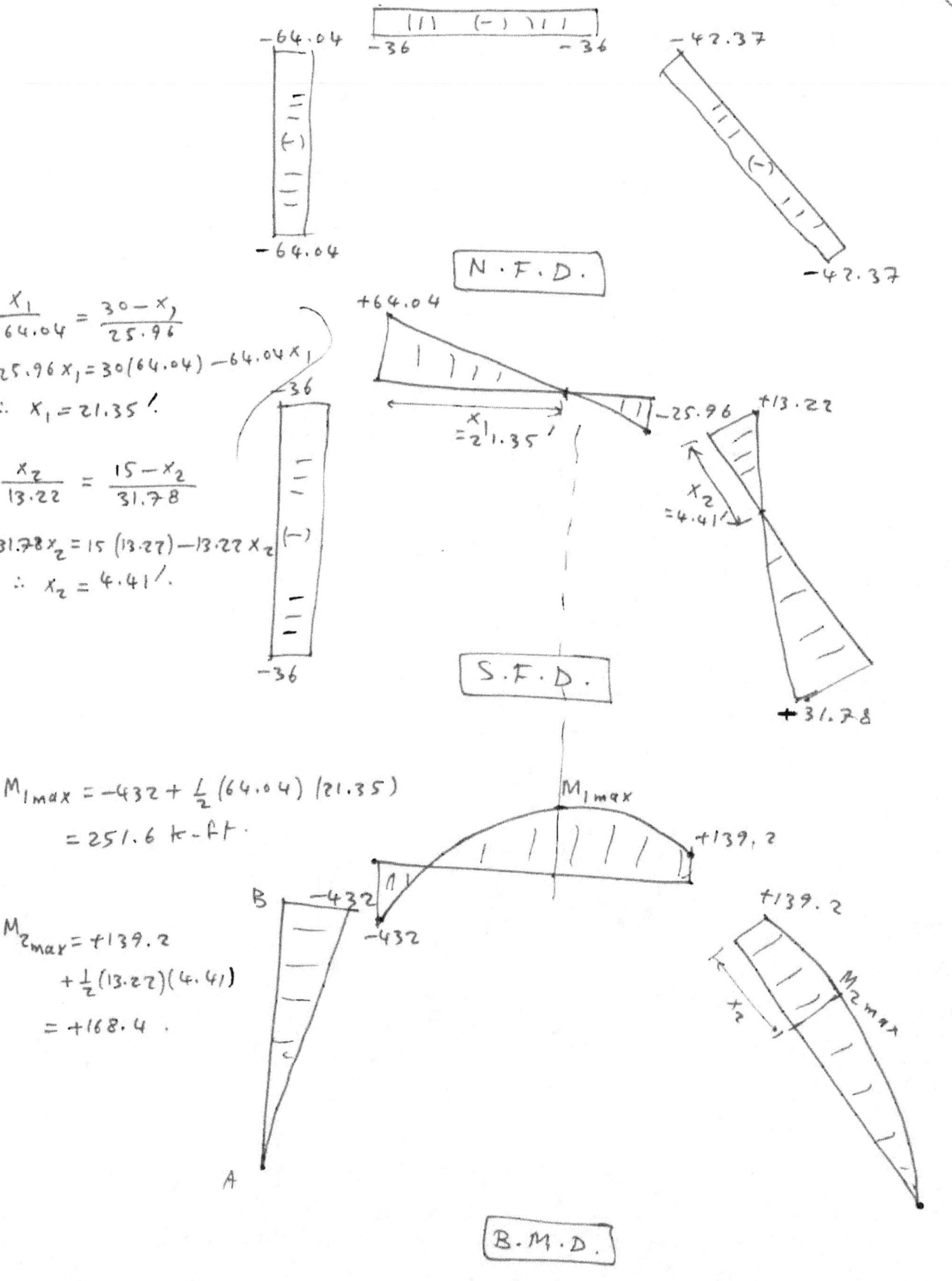

$$\frac{X_1}{64.04} = \frac{30 - X_1}{25.96}$$

$$25.96 X_1 = 30(64.04) - 64.04 X_1$$

$$\therefore X_1 = 21.35'.$$

$$\frac{X_2}{13.22} = \frac{15 - X_2}{31.78}$$

$$31.78 X_2 = 15(13.22) - 13.22 X_2$$

$$\therefore X_2 = 4.41'.$$

$$M_{1max} = -432 + \frac{1}{2}(64.04)(21.35)$$
$$= 251.6 \ k\text{-}ft.$$

$$M_{2max} = +139.2 + \frac{1}{2}(13.22)(4.41)$$
$$= +168.4.$$

Example 5.8 :

Construct the normal (axial), shear and moment diagrams for each member of the frame shown in the figure.

Solution :

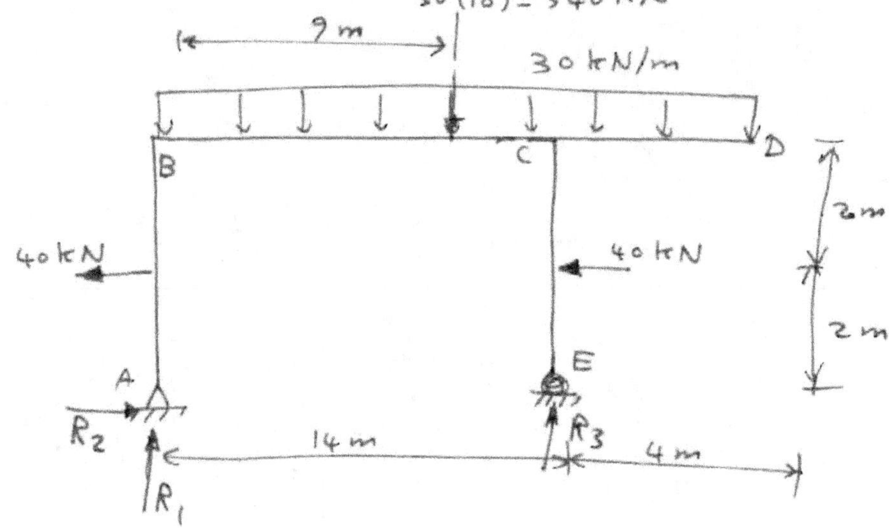

(1) Calculate the reactions :

$\xrightarrow{+} \Sigma F_x = 0:$ $R_2 - 40 - 40 = 0 \Rightarrow R_2 = 80 \ kN \longrightarrow$

$\overset{+}{\curvearrowright} \Sigma M_A = 0:$ $+40(2) + 40(2) - 540(9) + R_3(14) = 0$

$\Rightarrow R_3 = 335.7 \ kN \uparrow$

$+\uparrow \Sigma F_y = 0:$ $R_1 + 335.7 - 540 = 0 \Rightarrow R_1 = 204.3 \ kN \uparrow .$

(2) Consider each member of the frame separately :

Member AB :

$\overset{+}{\curvearrowright} \Sigma M_B = 0$

$M_B - 40(2) + 80(4) = 0$

$\therefore \ M_B = -240 \ kN \cdot m .$

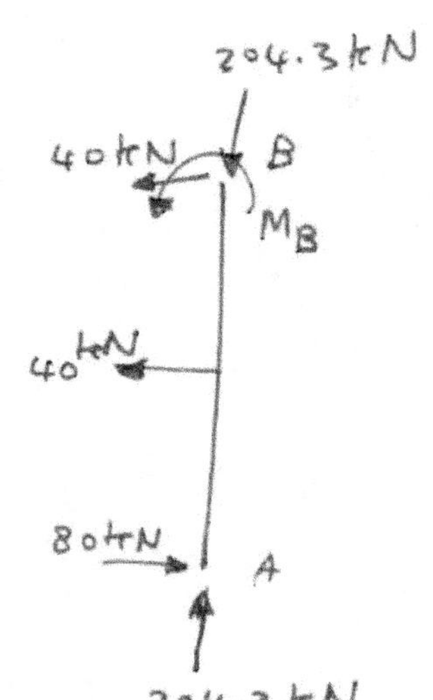

<u>Member BCD</u> :

$+\circlearrowleft \Sigma M_c = 0$:

$M_c + 540(5)$

$\quad - 204.3(14) + 240 = 0$

$\therefore M_c = -79.8 \ kN \cdot m$

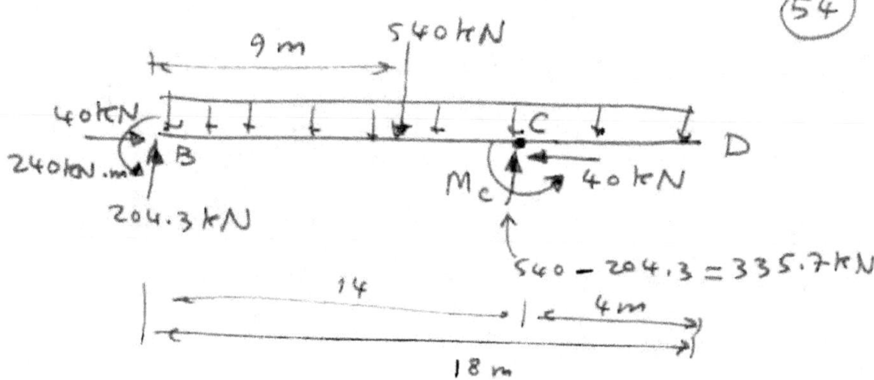

<u>Memer CE</u> :

<u>Check</u> :

$+\circlearrowleft \Sigma M_E = 0$:

$\quad 79.8 + 40(2) - 40(4) = 0$

$\qquad -0.2 = 0 \checkmark$

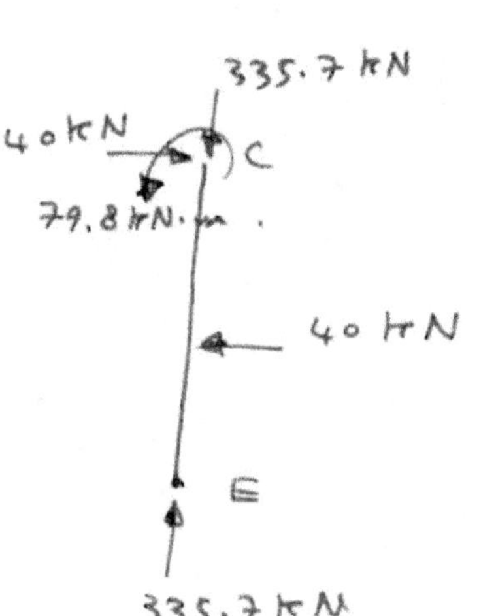

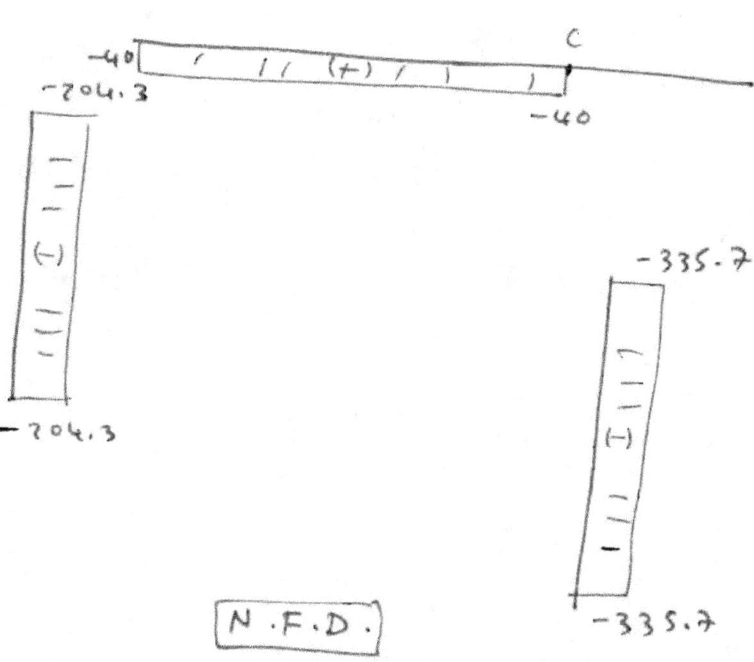

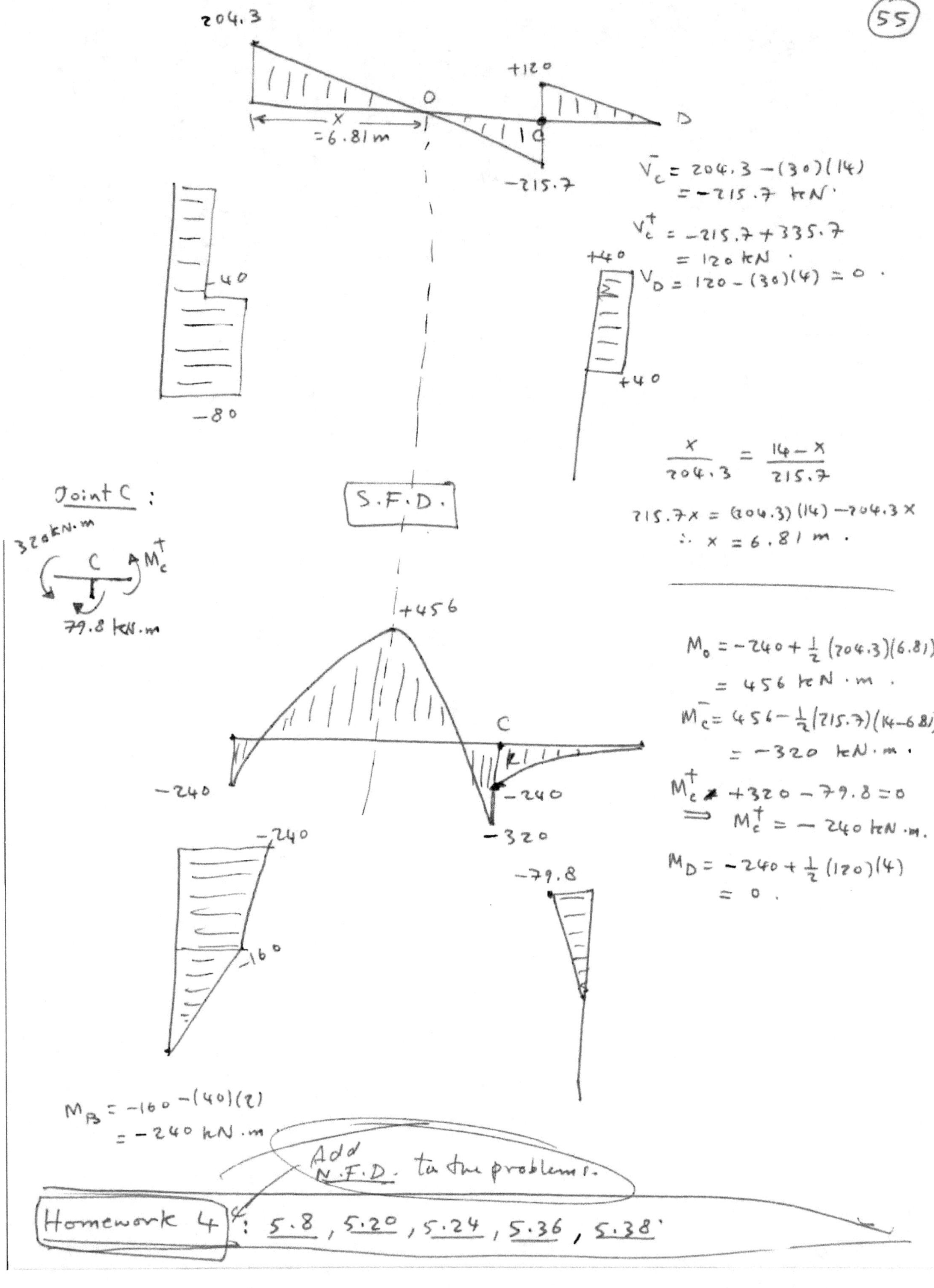

204.3

55

+120

O

D

x = 6.81 m

-215.7

$\bar{V_c} = 204.3 - (30)(14)$
$= -215.7$ kN

$V_c^+ = -215.7 + 335.7$
$= 120$ kN

$V_D = 120 - (30)(4) = 0$

+40

+40

-40

-80

$\dfrac{x}{204.3} = \dfrac{14-x}{215.7}$

$215.7x = (204.3)(14) - 204.3x$

$\therefore x = 6.81$ m

Joint C:

320kN·m

C M_c^+

79.8 kN·m

S.F.D.

+456

C

-240

-240

-240

-320

-79.8

-160

$M_0 = -240 + \frac{1}{2}(204.3)(6.81)$
$= 456$ kN·m

$\bar{M_c} = 456 - \frac{1}{2}(215.7)(14-6.81)$
$= -320$ kN·m

$M_c^+ = +320 - 79.8 = 0$
$\Rightarrow M_c^+ = -240$ kN·m

$M_D = -240 + \frac{1}{2}(120)(4)$
$= 0$

$M_B = -160 - (40)(2)$
$= -240$ kN·m

Add
N.F.D. to the problems.

Homework 4: 5.8, 5.20, 5.24, 5.36, 5.38

Example (1): (Inclined Beams):

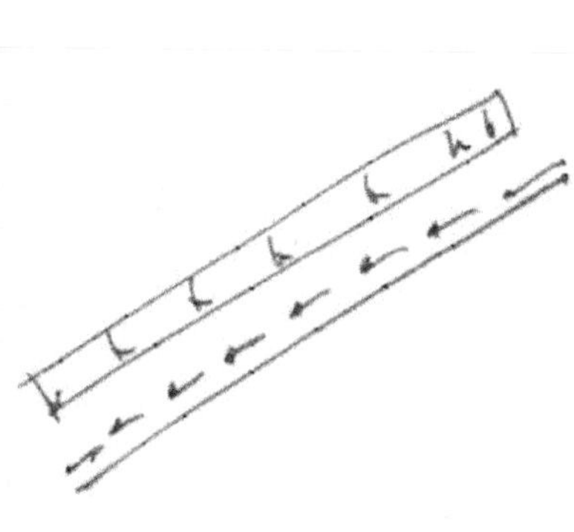

ω kN/m on horizontal projection

Total Load $= \omega L$.

Length of incline $= \dfrac{L}{\cos\alpha}$

Horizontal Load $= \omega L \sin\alpha$

Vertical Load $= \omega L \cos\alpha$

$\omega_H = \dfrac{\omega L \sin\alpha}{(L/\cos\alpha)} = \omega \sin\alpha \cos\alpha$

$\omega_V = \dfrac{\omega L \cos\alpha}{L/\cos\alpha} = \omega \cos^2\alpha$.

N.F.D.

S.F.D.

B.M.D.

$\cos\alpha = \dfrac{L}{2d}$

$d = \dfrac{L}{2\cos\alpha}$

$M_{max} = \text{area of triangle}$

$= \dfrac{1}{2}\left(\dfrac{\omega L}{2}\cos\alpha\right)\left(\dfrac{L}{2\cos\alpha}\right)$

$= \dfrac{\omega L^2}{8}$.

$V_0 = +\dfrac{\omega L}{2}\cos\alpha - (\omega\cos^2\alpha)\left(\dfrac{L}{2\cos\alpha}\right)$

$= \dfrac{\omega L}{2}\cos\alpha - \dfrac{\omega L}{2}\cos\alpha = 0$.

$N_0 = -\dfrac{\omega L}{2}\sin\alpha + (\omega\sin\alpha\cos\alpha)\left(\dfrac{L}{2\cos\alpha}\right)$

$= -\dfrac{\omega L}{2}\sin\alpha + \dfrac{\omega L}{2}\sin\alpha = 0$.

$N_x = -\dfrac{\omega L}{2}\sin\alpha + \cdots$

Chapter 6 : Deflections :
Differential Equation and Geometric Methods

6·1 Introduction :

* There are two reasons for calculating deflections :

(1) It is important to limit the size of the deflections in structures. Excessive deformations may prevent the structure from functioning properly.

(2) Deflections play a vital role in the analysis of indeterminate structures.

The analysis of an indeterminate structure requires that both internal forces and deformations be calculated simultaneously.

6·2 Elastic Force–Deformation Relationships :

* Assumptions :

(1) The material obeys Hooke's law, $\sigma = E\varepsilon$ (linear, elastic)

(2) The deformations are small compared to the original dimensions.

Example : Steel, $\sigma = 24\ ksi$, $E = 29 \times 10^3\ ksi$ (Young's modulus)

modulus of elasticity

$$\sigma = E\varepsilon$$
$$24 = 29 \times 10^3\ \varepsilon \implies \varepsilon = \frac{24}{29 \times 10^3} = 0.00083 \ \text{in./in.}$$
$$= \frac{83}{10,000}$$ (dimensionless).
$$= 0.083\ \% \ \text{(small deformation)}.$$

(1) Axially Loaded Bars :

Deflection $\delta = \dfrac{PL}{AE}$

$AE \equiv$ constant.

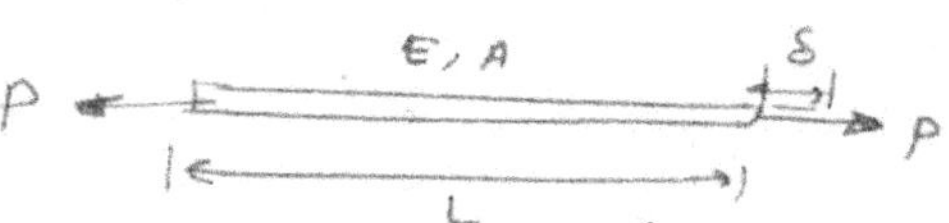

(2) Flexural Members : (Bending)

$$\frac{1}{r} = \frac{M}{EI}$$

$r \equiv$ radius of curvature

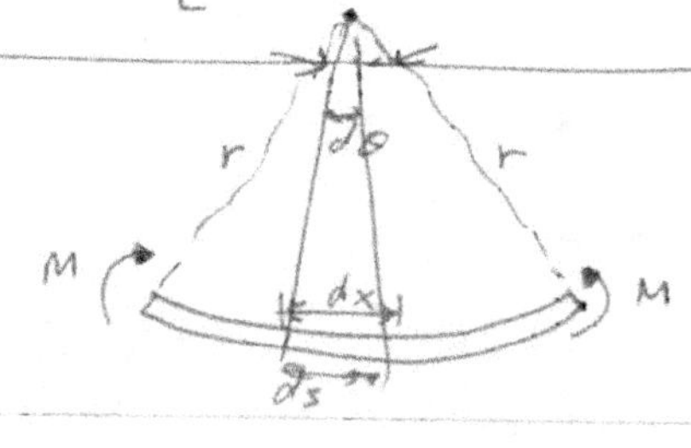

$I \equiv$ moment of inertia

From the figure: $ds = r\,d\theta$

$$\frac{1}{r} = \frac{d\theta}{ds}$$

$\Rightarrow \quad \dfrac{d\theta}{ds} = \dfrac{M}{EI}$, but $\theta = \dfrac{dy}{dx} =$ slope of the curve.

assume $\dfrac{d}{ds} = \dfrac{d}{dx}$

$$\Rightarrow \quad \boxed{\dfrac{d^2 y}{dx^2} = \dfrac{M}{EI}}$$

---

③ Torsional Members: 

ϕ : angle of twist

J : torsional constant

G : shear modulus, $\quad G = \dfrac{E}{2(1+\nu)}$, $\quad \nu$: Poisson's ratio

$$\phi = \frac{TL}{GJ} \quad , \quad T \ (\text{torsional or twisting moment}).$$

---

④ Shear Panels:

$$\delta = \frac{kVL}{AG}$$

$V \equiv$ shear force

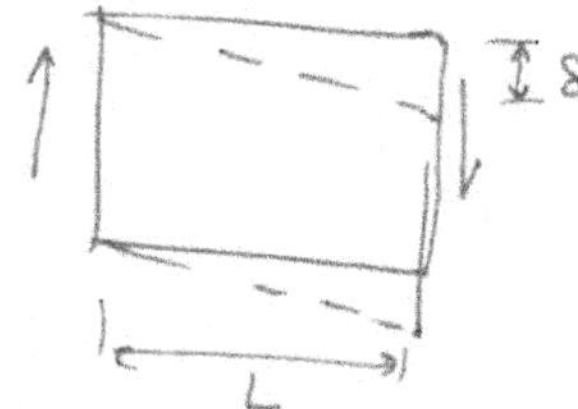

$k \equiv$ shape factor.

$k = 1.2$ for a rectangular section.

$k = 1.0$ for a wide flange member.

6.3 Direct Integration:

Example 6.1 :

Obtain an expression for the deflection of the cantilever beam shown in the figure. The stiffness EI of the beam is assumed to be constant.

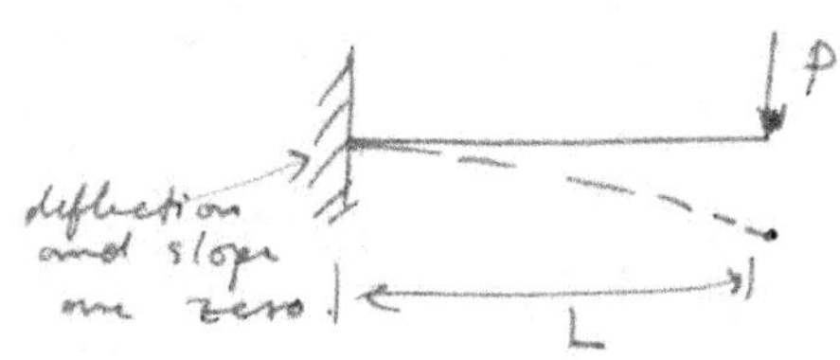

Solution :

$+\circlearrowleft \sum M_0 = 0$:

$$-M(x) - Px = 0$$

$$M(x) = -Px$$

$$M(x) = Px$$

$$\frac{d^2y}{dx^2} = \frac{M}{EI} = \frac{Px}{EI}$$

$$\frac{dy}{dx} = \int \frac{Px}{EI}\,dx = \frac{P}{EI}\frac{x^2}{2} + C_1$$

$$y(x) = \frac{P}{EI}\frac{x^3}{6} + C_1 x + C_2 .$$

Determine C_1 and C_2 from the boundary conditions:

at $x = L$, deflection $y = 0$.

at $x = L$, slope $\frac{dy}{dx} = 0$.

$$\frac{P}{6EI}L^3 + C_1 L + C_2 = 0 \quad\text{------- (1)}$$

$$\frac{P}{2EI}L^2 + C_1 = 0 \implies C_1 = -\frac{PL^2}{2EI}$$

$$\frac{PL^3}{6EI} - \frac{PL^3}{2EI} + C_2 = 0 \implies C_2 = \frac{PL^3}{3EI} .$$

$$\therefore \ y(x) = \frac{P}{6EI}x^3 - \frac{PL^2}{2EI}x + \frac{PL^3}{3EI}$$

$$\boxed{\ y(x) = \frac{P}{6EI}\left(x^3 - 3L^2 x + 2L^3\right)\ }\quad \text{Deflection equation.}$$

* The maximum deflection occurs at the free end, at $x = 0$.

$$\therefore \ y_{max} = y(0) = \frac{P}{6EI}\left(0 - 0 + 2L^3\right)$$

$$= \frac{PL^3}{3EI} .$$

Example 6.2 :

Obtain an expression for the deflection of the simply supported beam shown in the figure.

Solution :

(1) Find the reactions:

$\xrightarrow{+} \Sigma F_x = 0: \quad R_2 = 0.$

$\Sigma M_A = 0:$

$\quad R_3 L - Pa = 0 \Rightarrow R_3 = \dfrac{Pa}{L}.$

$+\uparrow \Sigma F_y = 0: \quad R_1 + \dfrac{Pa}{L} - P = 0 \Rightarrow R_1 = P - \dfrac{Pa}{L} = \dfrac{PL - Pa}{L}$

$$= \dfrac{P(L-a)}{L} = \dfrac{Pb}{L}.$$

(1) For $0 < x < a$; $\Sigma M_0 = 0, \circlearrowleft$

$\quad M(x) - \dfrac{Pb x}{L} = 0$

$\quad m(x) = \dfrac{Pb}{L} x$

$M(x), \quad 0 < x < a$

(2) For $a < x < L$:

$\quad m(x) = \dfrac{Pb x}{L} - P(x-a)$

$M(x), \quad a < x < L$

the moment is discontinuous at $x = a$:

① $0 < x < a$;

$$\dfrac{d^2 y}{dx^2} = \dfrac{M}{EI} = \dfrac{Pb x}{EIL}$$

$$\dfrac{dy}{dx} = \dfrac{Pb x^2}{2 EIL} + C_1$$

$$y_1(x) = \dfrac{Pb x^3}{6 EIL} + C_1 x + C_2 \quad\text{———— (1)}$$

② $a < x < L$, $\dfrac{d^2 y}{dx^2} = \dfrac{M}{EI} = \dfrac{1}{EI}\left[\dfrac{Pb x}{L} - P(x-a) \right]$

$$\dfrac{dy}{dx} = \dfrac{1}{EI}\left[\dfrac{Pb x^2}{2L} - \dfrac{P(x-a)^2}{2} + C_3 \right]$$

$$y_2(x) = \dfrac{1}{EI}\left[\dfrac{Pb x^3}{6L} - \dfrac{P(x-a)^3}{6} + C_3 x + C_4 \right] \quad\text{———— (2)}$$

Evaluate the **four** constants C_1, C_2, C_3, C_4 from the __boundary conditions__:

1. the deflection is zero at $x=0$, $y_1=0$.

2. the deflection is zero at $x=L$, $y=0$.

3. Condition of continuity, at $x=a$, $y_1=y_2$. (deflection)

4. Condition of continuity, at $x=a$, $\dfrac{dy_1}{dx}=\dfrac{dy_2}{dx}$. (slope)

(1) $\quad 0 = 0 + 0 + C_2 \implies C_2 = 0$.

(2) $\quad 0 = \dfrac{1}{EI}\left[\dfrac{Pb L^3}{6L} - P\dfrac{(L-a)^3}{6} + C_3 L + C_4\right]$

$\implies \dfrac{1}{6}Pb L^2 - P\dfrac{b^3}{6} + C_3 L + C_4 = 0 \quad\text{———}\boxed{A}$

(3) $\quad \dfrac{Pba^3}{6EIL} + C_1 a = \dfrac{1}{EI}\left[\dfrac{Pba^3}{6L} - 0 + C_3 a + C_4\right]$

$\dfrac{Pba^3}{6EIL} + C_1 a = \dfrac{1}{EI}\left[\dfrac{Pba^3}{6L} + C_3 a + C_4\right]$

$\dfrac{Pba^3}{6EIL} + C_1 a = \dfrac{Pba^3}{6EIL} + \dfrac{1}{EI}(C_3 a + C_4)$

$\implies C_1 a = \dfrac{1}{EI}(C_3 a + C_4) \quad\text{———}\boxed{B}$

(4) $\quad \dfrac{Pba^2}{2EIL} + C_1 = \dfrac{1}{EI}\left[\dfrac{Pba^2}{2L} - 0 + C_3\right]$

$\boxed{C_1 = C_3} \quad\text{———}\boxed{C}$

Substitute in eq. $\boxed{B}$:

$C_3 a = \dfrac{1}{EI}(C_3 a + C_4)$

$EI\, a\, C_3 - C_3 a = C_4$

$C_3\, a\,(EI - 1) = C_4$

Substitute in eq. $\boxed{A}$:

$\dfrac{Pb L^2}{6} - \dfrac{Pb^3}{6} + C_3 L + C_3 a(EI - 1) = 0$

$\dfrac{1}{6}Pb(L^2 - b^2) = C_3 a(1 - EI) - C_3 L$

$\dfrac{1}{6}Pb(L^2 - b^2) = C_3\left[a - a\,EI - L\right]$

$$\therefore \quad \frac{1}{6} Pb(L^2 - b^2) = C_3\left[-aEI - b\right] \quad \text{since} \quad a - L = -(L-a) = -b$$ (61)

$$\frac{1}{6} Pb(L^2 - b^2) = -C_3(b + aEI)$$

$$\therefore \quad C_3 = \frac{-Pb(L^2 - b^2)}{6(b + aEI)}$$

$$\therefore \quad C_1 = C_3 = \frac{-Pb(L^2 - b^2)}{6(b + aEI)}$$

$$C_4 = C_3 a(EI - 1) = \frac{-Pab(L^2 - b^2)(EI - 1)}{6(b + aEI)} \qquad \text{Final Answer is in the book.}$$

Example 6.3:

Obtain an expression for the deflection of the beam in the previous example using __singularity functions__.

__Solution :__

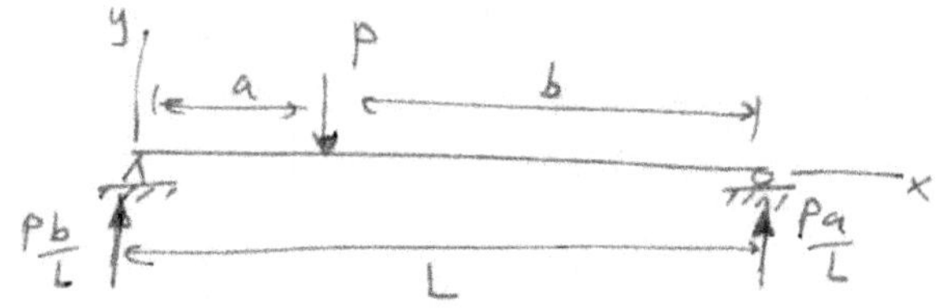

$$M(x) = \frac{Pbx}{L} - P\langle x - a \rangle$$

The function $\langle x - a \rangle$ is called a __singularity function__, defined by:

$$\langle x - a \rangle = \begin{cases} 0, & x - a < 0, \text{ i.e. } x < a \\ x - a, & x - a > 0, \text{ i.e. } x > a \end{cases}$$

$$\frac{d^2 y}{dx^2} = \frac{M}{EI} = \frac{1}{EI}\left[\frac{Pbx}{L} - P\langle x - a \rangle\right]$$

$$\frac{dy}{dx} = \frac{1}{EI}\left[\frac{Pbx^2}{2L} - \frac{P}{2}\langle x - a \rangle^2 + C_1\right]$$

$$y(x) = \frac{1}{EI}\left[\frac{Pbx^3}{6L} - \frac{P}{6}\langle x - a \rangle^3 + C_1 x + C_2\right]$$

We only have two boundary conditions:

$$y = 0 \quad \text{at} \quad x = L \text{ and } x = 0.$$

(1) $\quad 0 = \dfrac{1}{EI}\left[0 - 0 + 0 + C_2\right] \implies C_2 = 0.$

(2) $\quad 0 = \dfrac{1}{EI}\left[\dfrac{PbL^3}{6L} - \dfrac{P\langle L - a \rangle^3}{6} + C_1 L\right]$

but $\langle L-a \rangle = \langle b \rangle = b$, since $b > 0$.

$$\therefore \quad \frac{PbL^2}{6} - \frac{Pb^3}{6} + C_1 L = 0$$

$$\frac{Pb}{6}(L^2 - b^2) = -C_1 L$$

$$\implies \quad C_1 = -\frac{Pb}{6L}(L^2 - b^2).$$

$$\therefore \quad y(x) = \frac{1}{EI}\left[\frac{Pbx^3}{6L} - \frac{P}{6}\langle x-a\rangle^3 - \frac{Pb}{6L}(L^2-b^2)x\right]$$

$$\boxed{y(x) = \frac{1}{EI}\left[\frac{Pbx}{6L}(x^2+b^2-L^2) - \frac{1}{6}P\langle x-a\rangle^3\right].}$$

6.4 Moment-Area Theorems:

* useful in simple beams.

* it can only be used to determine the _slope_ or _deflection_ at a specific point along the member.

* it uses the bending moment diagram.

$$\frac{d\theta}{ds} = \frac{M}{EI} \quad \text{or} \left(\frac{d^2y}{dx^2}\right)$$

For small deformations,
$$d\theta \approx dx$$

$$\boxed{d\theta = \frac{M}{EI}dx}$$

The change in slope is equal to the area of the $\frac{M}{EI}$ diagram.

$$\theta_{CD} = \int_c^b \frac{M}{EI}dx \quad .$$

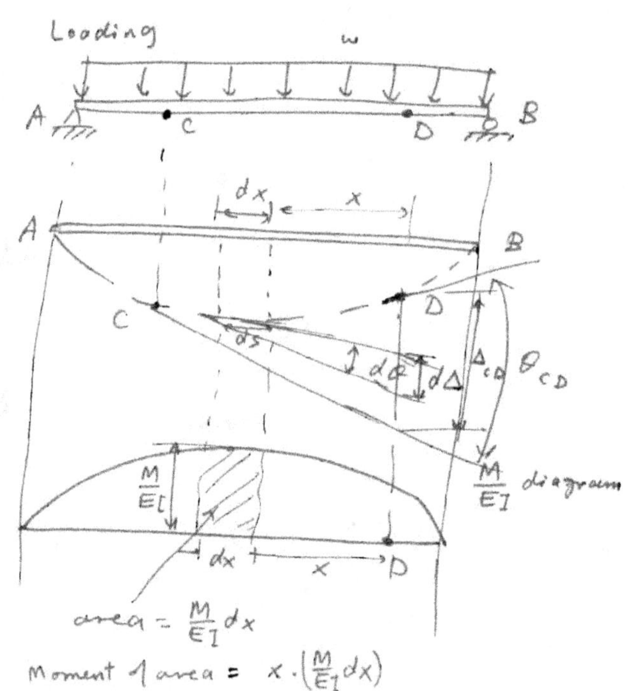

Theorem 1:

The change in slope between two points on the defl.
of a beam is equal to the area of the $\frac{M}{EI}$ diagram between
these points.

From the figure: $\quad d\Delta = x\, d\theta$

$$\text{but}\quad d\theta = \frac{M}{EI}dx$$

$$\Rightarrow \quad d\Delta = x\left(\frac{M}{EI}dx\right)$$

the vertical distance between the tangents is equal to the
moment of the $\frac{M}{EI}$ diagram about point D.

$$\Delta_{CD} = \int_{C}^{D} x\left(\frac{M}{EI}dx\right).$$

Theorem 2:

The vertical distance Δ, from a point D on the deflection
curve of a beam to a tangent drawn to some other point C,
is equal to the moment of the $\frac{M}{EI}$ diagram between C and D
about D.

Notes:

(1) The theorem does **not** give a beam deflection. Instead, it
gives the vertical distance from one point on a beam to a
tangent drawn to some other point.

(2) The moment of the $\frac{M}{EI}$ diagram is always taken about the
point on the beam at which the vertical distance is measured.

6.5 Application of the Moment-Area Method:

Notes: Areas and Centroids:

① Triangle
$$A = \frac{1}{2}bh$$
$$\bar{x} = \frac{2}{3}b \quad.$$

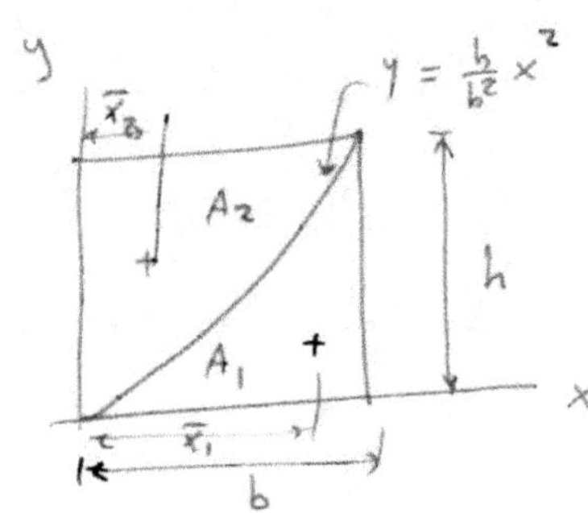

$$A_1 = \frac{1}{3}bh \quad , \quad \bar{x}_1 = \frac{3}{4}b$$

$$A_2 = \frac{2}{3}bh \quad , \quad \bar{x}_2 = \frac{3}{8}b \quad .$$

② Parabola.

③ Cubic :

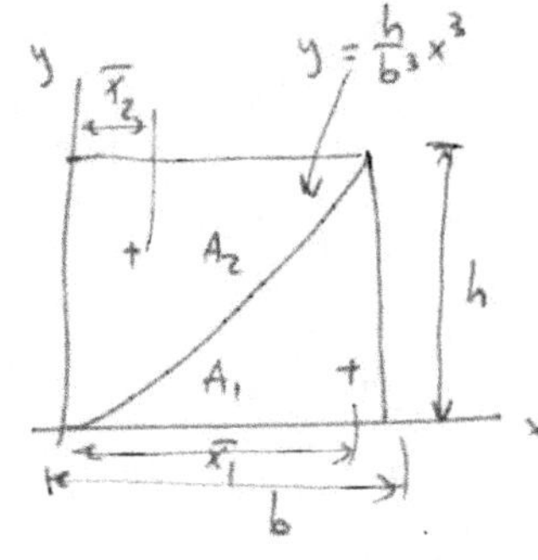

$$A_1 = \frac{1}{4}bh \quad , \quad \bar{x}_1 = \frac{4}{5}b$$

$$A_2 = \frac{3}{4}bh \quad , \quad \bar{x}_2 = \frac{2}{5}b \quad .$$

Example 6.4 :

Determine the slope and deflection at the free end of the cantilever beam shown in the figure. The beam is assumed to possess a constant stiffness EI.

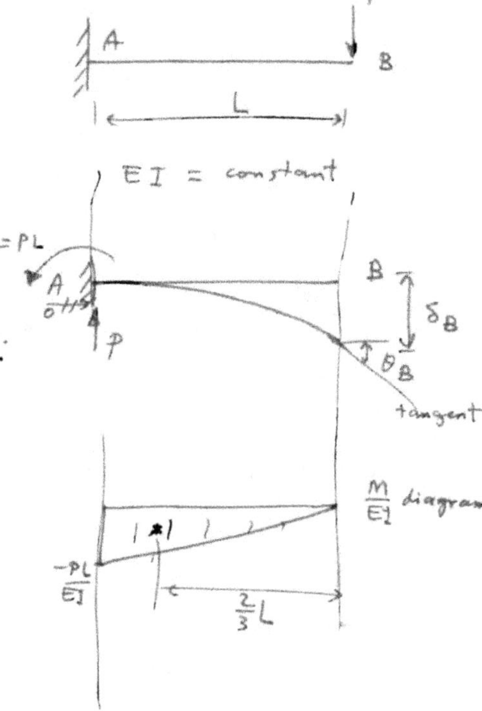

Solution :

For a fixed end ⟹ a horizontal tangent.

According to theorem 1 ,

$$\Delta\theta_{AB} = \int \frac{M}{EI}\, dx$$

$$= \text{area under } \frac{M}{EI} \text{ diagram}$$

$$= \frac{1}{2}\left(\frac{PL}{EI}\right)L = \frac{1}{2}\frac{PL^2}{EI}$$

$$\theta_B - \theta_A = \frac{PL^2}{2EI} \quad , \quad \text{but } \boxed{\theta_A = 0}$$

$$\therefore \quad \theta_B = \frac{PL^2}{2EI} \quad .$$

According to theorem 2,

$$d\Delta_{AB} = x\left(\frac{M}{EI}dx\right)$$

$$= \text{moment of area of } \frac{M}{EI} \text{ diagram about } B$$

$$= \frac{1}{2}\left(\frac{PL}{EI}\right)(L)\cdot\left(\frac{2}{3}L\right)$$

$$= \frac{PL^3}{3EI}$$

$$\delta_B - \overset{0}{\cancel{\delta_A}} = \frac{PL^3}{3EI}$$

$$\delta_B = \frac{PL^3}{3EI}$$

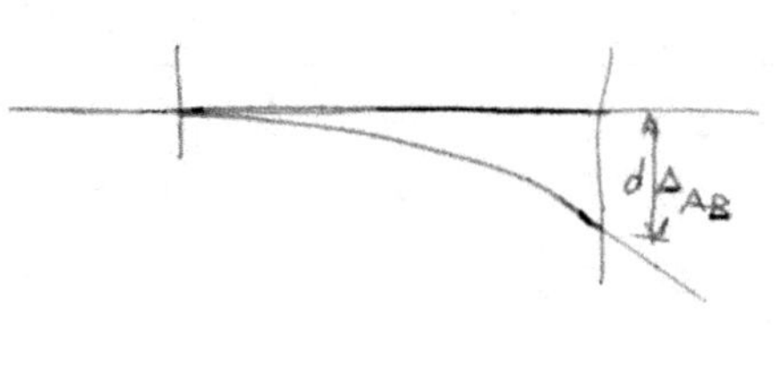

Example 6.5 :

For the simply-supported beam shown in the figure, determine the deflection at midspan and the slope at A.

Solution :

The slope at B is <u>zero</u>, (horizontal tangent), from symmetry.

$$\Delta\theta_{AB} = \text{area of } \frac{M}{EI} \text{ diagram}$$

$$= \frac{2}{3}\left(\frac{L}{2}\right)\left(\frac{wL^2}{8EI}\right)$$

$$= \frac{wL^3}{24EI}$$

$$\overset{0}{\cancel{\theta_B}} - \theta_A = \frac{wL^3}{24EI}$$

$$\theta_A = \left|-\frac{wL^3}{24EI}\right| = \frac{wL^3}{24EI}$$

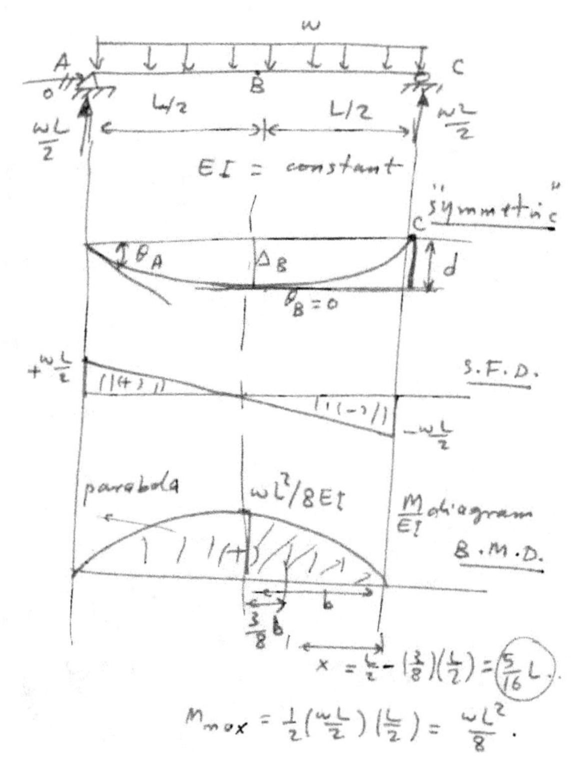

From theorem 2,

$$d = \text{moment of area of } \frac{M}{EI} \text{ diagram about } C$$

$$= \frac{2}{3}\left(\frac{L}{2}\right)\left(\frac{wL^2}{8EI}\right)\cdot\frac{5}{16}L = \frac{5wL^4}{384EI}$$

Example 6.6 :

It is required to determine the deflection at point B for the cantilever beam shown in the figure. The stiffness of the beam is equal to EI between B and C and $2EI$ between A and B.

Solution :

From Theorem 2,

δ_B = moment of the $\frac{M}{EI}$ diagram between A and B about B.

$$= \frac{PL}{4EI}\left(\frac{L}{2}\right)\cdot\left(\frac{L}{4}\right)$$

$$+ \frac{1}{2}\left(\frac{PL}{2EI}\right)\left(\frac{L}{2}\right)\cdot\left(\frac{1}{3}L\right)$$

$$= \frac{PL^3}{32EI} + \frac{PL^3}{24EI}$$

$$= \frac{7PL^3}{96EI}.$$

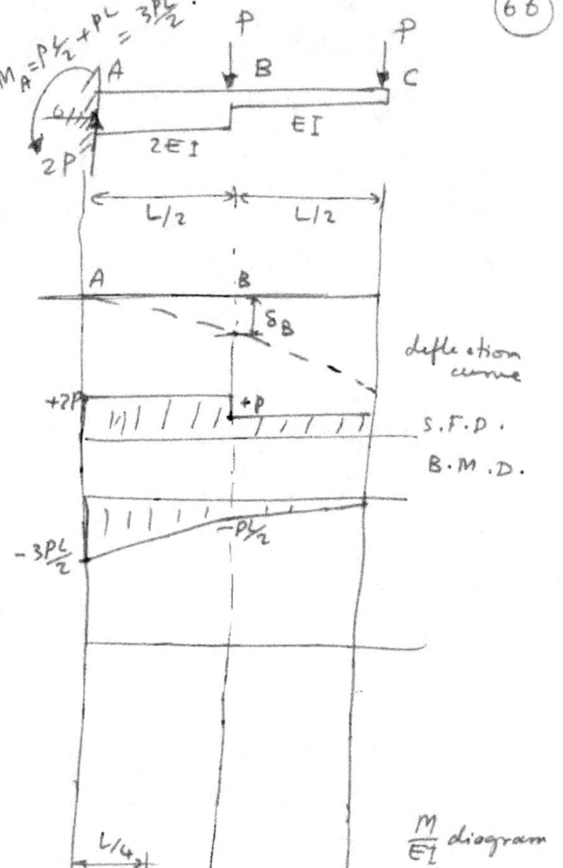

Example 6.7 :

Determine the slope at A and the deflection at B for the simply-supported beam shown in the figure.

$$E = 200 \times 10^6 \ kN/m^2$$

$$I = 100 \times 10^6 \ mm^4$$

$$EI = \left(200 \times 10^6 \ \frac{kN}{m^2}\right)\left(100 \times 10^6 \times (10^{-3})^4 \ m^2\right)$$

$$= 20,000 \ kN \cdot m^2.$$

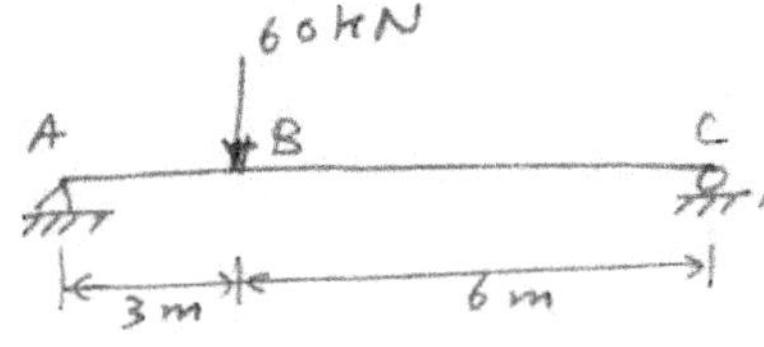

<u>Solution</u> : $EI = 20,000$ kN·m²

(1) Find the reactions :

$$R_{Ay} = \frac{60 \times 6}{9} = 40 \text{ kN}.$$

$$R_C = \frac{60 \times 3}{9} = 20 \text{ kN}.$$

$$R_{Ax} = 0.$$

The beam and the deflection curve are <u>not symmetric</u>.

(2) From similar triangles,

$$\frac{\delta_B + d_2}{3} = \frac{d_1}{9}$$

$$\Rightarrow \quad \delta_B = \tfrac{1}{3}d_1 - d_2$$

d_1 and d_2 are determined using theorem 2 :

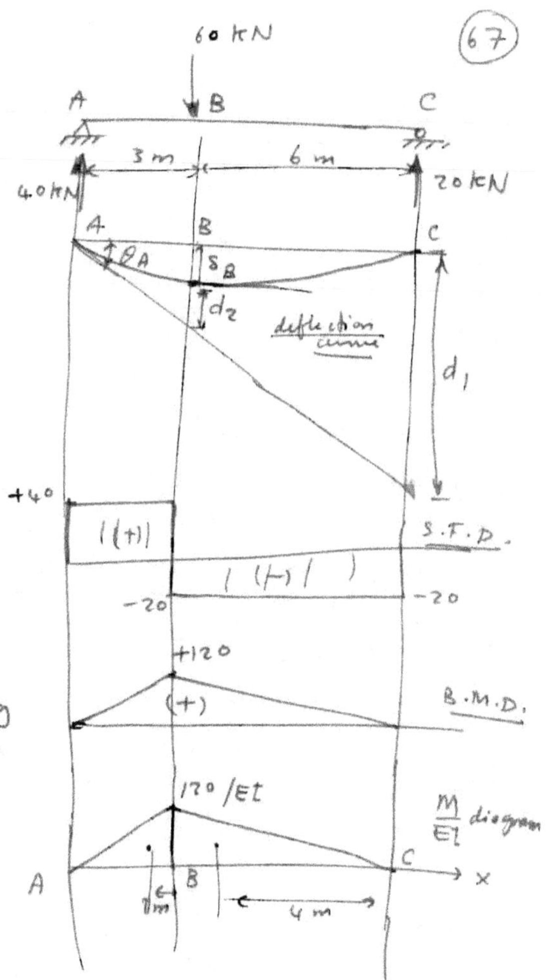

$d_1 =$ moment of $\frac{M}{EI}$ diagram between A and C about C

$$= \tfrac{1}{2}\left(\frac{120}{EI}\right)(3)(7) + \tfrac{1}{2}\left(\frac{120}{EI}\right)(6)(4)$$

$$= \frac{1260}{EI} + \frac{1440}{EI}$$

$$= \frac{2700}{EI}.$$

$d_2 =$ moment of $\frac{M}{EI}$ diagram between A and B about B

$$= \tfrac{1}{2}\left(\frac{120}{EI}\right)(3)(1) \quad = \frac{180}{EI}.$$

$$\therefore \; \delta_B = \tfrac{1}{3}\left(\frac{2700}{EI}\right) - \frac{180}{EI} = \frac{720}{EI} = \frac{720}{20,000} = 0.036 \text{ m} = \underline{3.6 \text{ cm}}.$$

Since the slopes and deformations are small, $\boxed{\tan\theta_A \approx \theta_A}$

$$\theta_A \approx \tan\theta_A = \frac{d_1}{9} = \frac{2700}{9EI} = \frac{2700}{9(20,000)} = 0.015 \text{ rad}.$$

Example 6.8 :
Determine the deflection at C
for the beam shown in the
figure.

Solution :

(1) Find the reactions:

$+\curvearrowright \Sigma m_A = 0$:

$$R_B(L) - P\left(3\tfrac{L}{2}\right) = 0$$

$$R_B = 3P/2 \uparrow .$$

$+\uparrow \Sigma F_y = 0$:

$$R_{Ay} + \tfrac{3P}{2} - P = 0$$

$$R_{Ay} = -P/2 .$$

$$\therefore R_{Ay} = P/2 \downarrow .$$

$\xrightarrow{+} \Sigma F_x = 0 , \quad R_{Ax} = 0 .$

(2) $\quad \delta_c = d_3 - d_2 .$

We can determine d_1 and d_2
from theorem 2 :

$d_1 =$ moment of $\frac{M}{EI}$ diagram between
  A and B about B

$$= \tfrac{1}{2}\left(\frac{PL}{2EI}\right)(L)\cdot\left(\tfrac{1}{3}L\right)$$

$$= \frac{PL^3}{12EI}$$

$d_3 =$ moment of $\frac{M}{EI}$ diagram between A and C about C

$$= \tfrac{1}{2}\left(\frac{PL}{2EI}\right)(L)\cdot\left(\tfrac{L}{2}+\tfrac{L}{3}\right) + \tfrac{1}{2}\left(\frac{PL}{2EI}\right)\left(\tfrac{L}{2}\right)\left(\tfrac{L}{3}\right)$$

$$= \frac{5PL^3}{24EI} + \frac{PL^3}{24EI}$$

$$= \frac{PL^3}{4EI} .$$

From similar triangles, $\quad \dfrac{d_1}{d_2} = \dfrac{L}{3\frac{L}{2}} = \dfrac{2}{3} \implies d_2 = \tfrac{3}{2}d_1 = \tfrac{3}{2}\dfrac{PL^3}{12EI}$

$$= \frac{PL^3}{8EI}$$

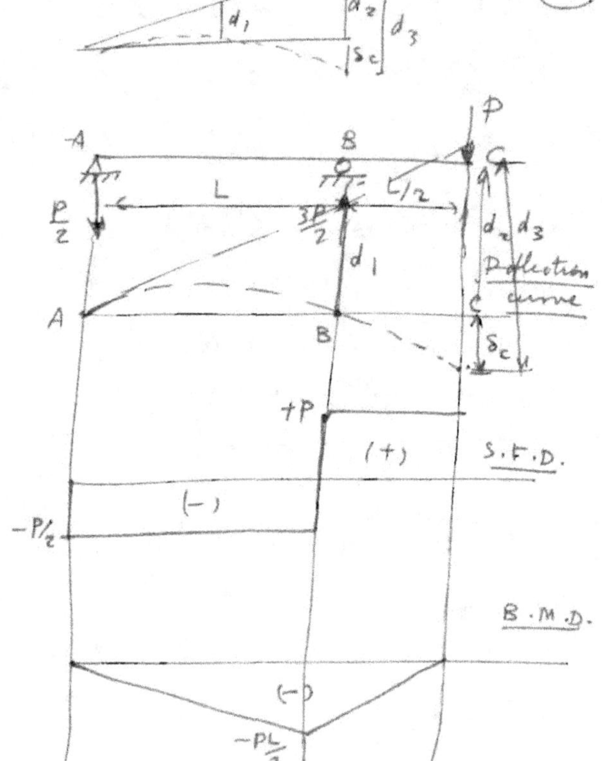

$$\delta_c = d_3 - d_2$$
$$= \frac{PL^3}{4EI} - \frac{PL^3}{8EI}$$
$$= \frac{PL^3}{8EI} .$$

Example 6.9 :

Determine the deflection at B for the beam shown in the figure.

Solution :

Use the principle of superposition.

$$E = 29 \times 10^3 \text{ ksi}$$
$$I = 150 \text{ in}^4$$
$$EI = (29 \times 10^3)(150)$$
$$= 4.35 \times 10^6 \text{ k-in}^2.$$

From theorem 2,
For the distributed load:

$$d_1 = \frac{2}{3}\left(\frac{100}{EI}\right)(10)\left(\frac{5}{8}\right)(10) = \frac{4667}{EI}$$

$$\delta_{B_1} = d_1 = \frac{4167}{EI} .$$

For the concentrated load:

$$d_2 = \frac{1}{2}\left(\frac{45}{EI}\right)(5)\left(\frac{2}{3}\right)(5)$$
$$+ \frac{1}{2}\left(\frac{45}{EI}\right)(15)(5+5)$$
$$= \frac{3750}{EI} .$$

From similar triangles
$$\frac{y}{45} = \frac{10}{15}$$
$$\Rightarrow y = 30/EI .$$

$$d_3 = \frac{1}{2}\left(\frac{30}{EI}\right)(10)\left(\frac{1}{3}(10)\right)$$
$$= \frac{500}{EI}$$

$$\delta_{B_2} = \frac{1}{2}d_2 - d_3 = \frac{1375}{EI} .$$

by superposition :
$$\delta_B = \delta_{B_1} + \delta_{B_2} = \frac{4167}{EI} + \frac{1375}{EI} = \frac{5542}{EI}$$

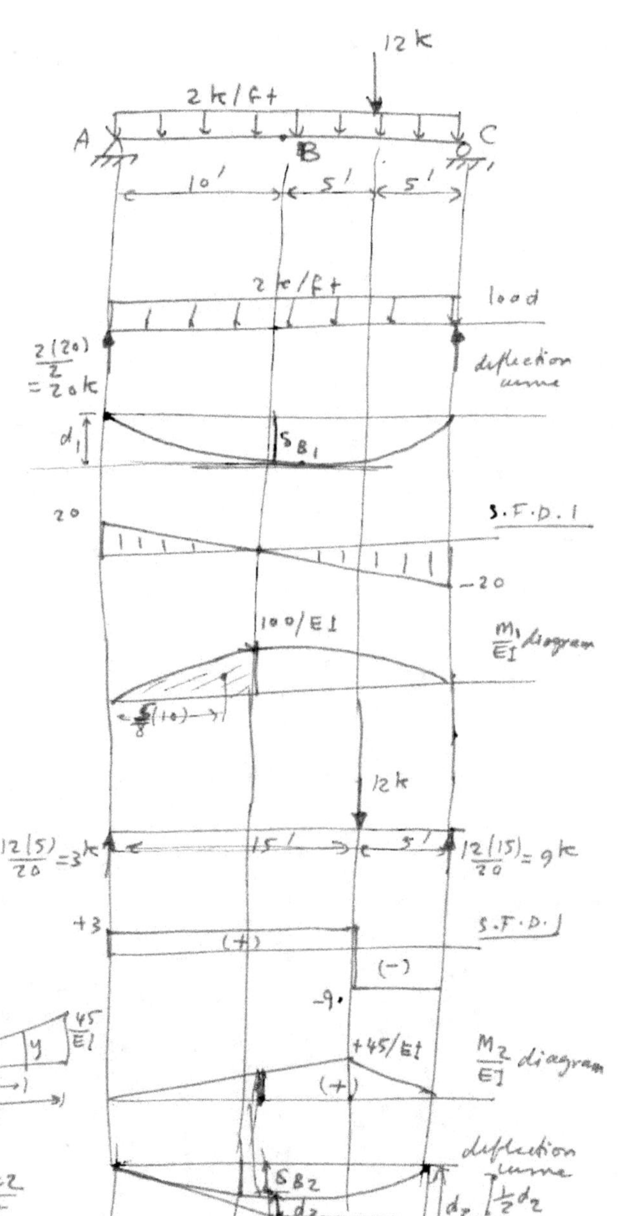

$$\therefore \; \delta_B = \frac{5542}{\left[\frac{4.35 \times 10^6}{(12^2)}\right]} \longrightarrow = 0.183 \; ft = 2.20 \; in.$$

Example 6.10 :

Determine the deflection at C for the beam shown in the figure.

Solution : $E = 200 \times 10^6 \; kN/m^2$.

$\quad\quad\quad\quad I = 200 \times 10^6 \; mm^4$.

$$\delta_c = \tfrac{1}{2} d_1 - d_2 .$$

Calculate the reactions :

$\circlearrowleft \Sigma F_x = 0 : R_2 = 0 :$

$\circlearrowleft \Sigma M_A = 0 :$

$\quad\quad R_3 (12) - 40(3) - 60(9) = 0$

$\quad\quad\quad R_3 = 55 \; kN \uparrow$

$+\uparrow \Sigma F_y = 0 :$

$\quad\quad R_1 + 55 - 40 - 60 = 0$

$\quad\quad\quad R_1 = 45 \; kN \uparrow$

$\underline{V_D = 5 - (10)(60) = -55 \; kN .}$

$M_B = 45(3) = 135 \; kN \cdot m$

$M_c = 135 + (5)(3) = 150 \; kN \cdot m .$

$\quad \dfrac{x}{6-x} = \dfrac{5}{55} \Longrightarrow 55x = 30 - 5x$

$\quad\quad\quad\quad\quad 60x = 30$

$\quad\quad\quad\quad\quad x = 0.5 \; m .$

$M_0 = 150 + \tfrac{1}{2}(5)(\tfrac{1}{2}) = 151.25 \; kN \cdot m .$

$M_D = 151.25 - \tfrac{1}{2}(55)(5.5) = 0 .$

Divide the B.M.D. into segments :

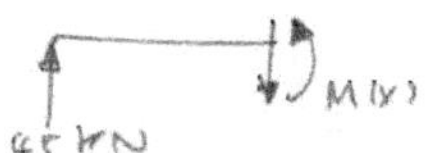

$M(x) = 45x , \quad 0 < x < 3$

$M(x) = 45x - 40(x-3) , \quad 3 < x < 6$

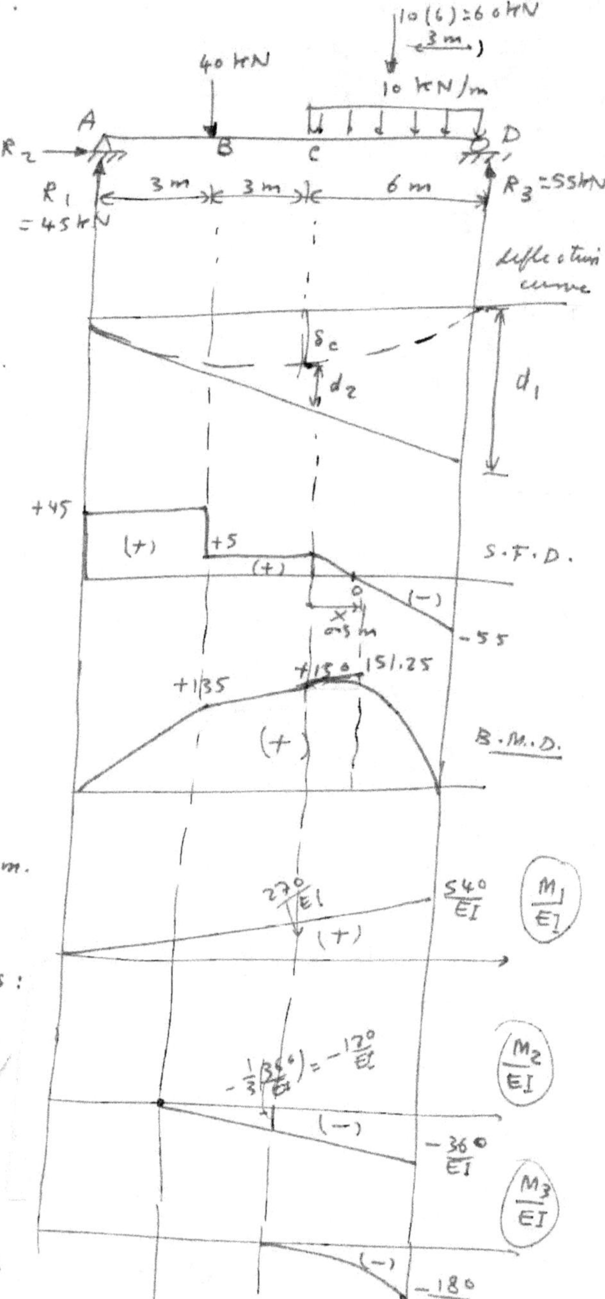

$$M(x) = 45x - 40(x-3) - 10\frac{(x-6)^2}{2}, \qquad 6 < x < 12$$

$$d_1 = \frac{1}{2}\left(\frac{540}{EI}\right)(12)\left(\frac{1}{3}\right)(12) - \frac{1}{2}\left(\frac{360}{EI}\right)(9)\left(\frac{1}{3}\right)(9) - \frac{1}{3}\left(\frac{180}{EI}\right)(6)\left(\frac{1}{4}\right)(6)$$

$$= \frac{7560}{EI}.$$

$$d_2 = \frac{1}{2}\left(\frac{270}{EI}\right)(6)\left(\frac{1}{3}\right)(6) - \frac{1}{2}\left(\frac{120}{EI}\right)(3)\left(\frac{1}{3}\right)(3) = \frac{1440}{EI}.$$

$$\therefore \; \delta_c = \frac{1}{2}d_1 - d_2 = \frac{1}{2}\left(\frac{7560}{EI}\right) - \frac{1440}{EI} = \frac{2340}{EI}.$$

$$\therefore \; \delta_c = \frac{2340}{(200\times10^6)(200\times10^6)\times(10^{-3})^4} = 0.0585 \; m$$

$$= 5.85 \; cm.$$

6.6 Conjugate–Beam Method:

$$+\uparrow \; \Sigma F_y = 0:$$

$$V - (V + dV) - w\,dx = 0$$

$$dV = -w\,dx.$$

$$\boxed{V = -\int w\,dx}$$

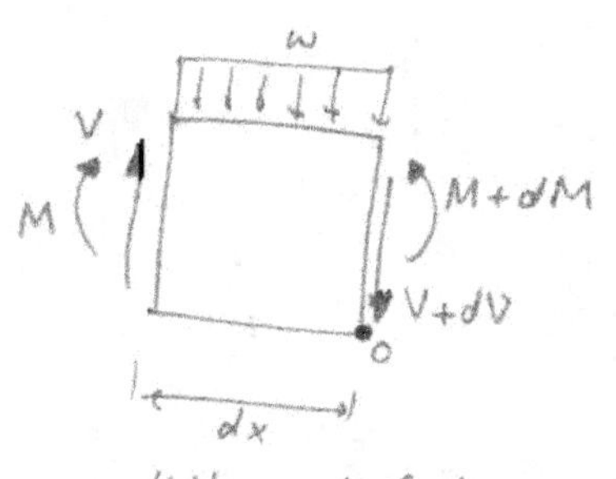

differential beam element

$$+\circlearrowleft \; \Sigma M_0 = 0:$$

$$(M + dM) - M - V\,dx + w(dx)\left(\frac{dx}{2}\right) = 0$$

$$dM - V\,dx + \frac{1}{2}w(dx)^2 = 0$$

very small (neglected).

$$dM = V\,dx$$

$$\boxed{M = \int V\,dx}$$

$$\therefore \ M = \int \left(-\int w\,dx\right)dx \quad \Rightarrow \quad \boxed{M = \iint w\,dx\,dx} \quad \text{Double Integration}$$

Similarly,

$$\frac{d^2 y}{dx^2} = \frac{M}{EI}$$

$$\Rightarrow \ \text{slope} \equiv \frac{dy}{dx} = \int \frac{M}{EI}\,dx$$

$$\text{deflection} \quad \boxed{y = \iint \frac{M}{EI}\,dx\,dx} \qquad (\text{Double Integration}).$$

Use $\left(\dfrac{M}{EI}\right)$ as a load on a new beam, (fictitious) called a __conjugate beam__.

* Introduce a new (fictitious) beam called the __conjugate beam__.

* It has : (1) The same length as the original beam.

(2) the load is the $\frac{M}{EI}$ diagram (always a distributed load).

> ① The __shear__ at any section of the conjugate beam is equal to the __slope__ at that section in the real beam.
>
> ② The __moment__ at any section of the conjugate beam is equal to the __deflection__ of the corresponding section of the real beam.

Support Conditions :

Real Beam	Conjugate Beam

① Fixed end $y = \frac{dy}{dx} = 0$ Free end $M = V = 0$

② Hinge $\frac{dy}{dx} \neq 0$, $y = 0$ Hinge $V \neq 0$, $M = 0$

③ Free end $y \neq 0$, $\frac{dy}{dx} \neq 0$ Fixed end $M \neq 0$, $V \neq 0$.

(4)   <u>Real Beam</u>

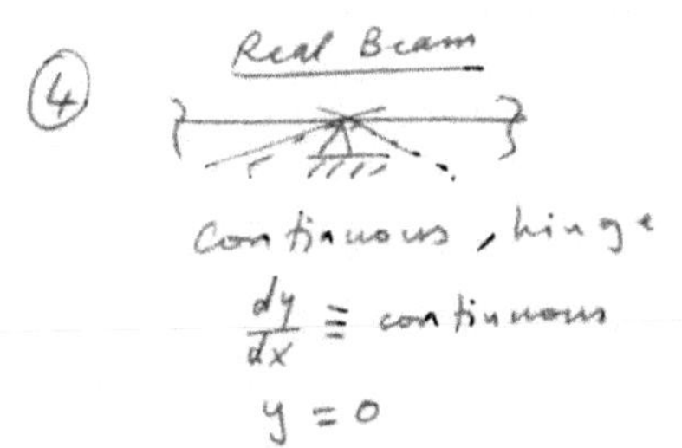

<u>Conjugate Beam</u> internal hinge (73)

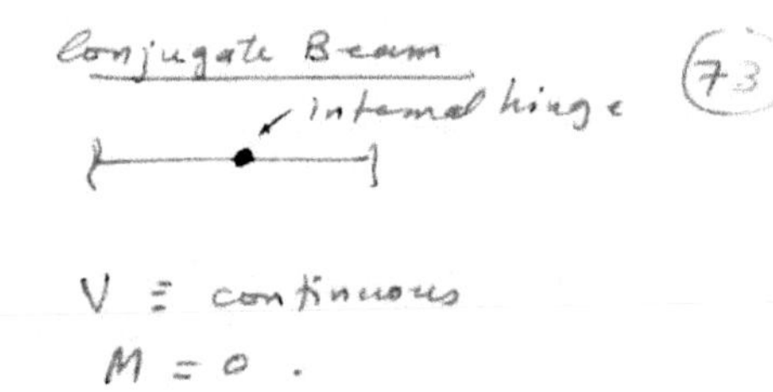

Continuous, hinge

$$\frac{dy}{dx} \equiv \text{continuous}$$

$$y = 0$$

$V \equiv$ continuous

$M = 0$.

---

* The conjugate beam method is used for simple beams.

6.7 Application of the Conjugate Beam Method:

<u>Example 6.11</u> :

Determine the slope at A and the deflection at B for the beam shown in the figure.

$$\theta_A \equiv \frac{dy}{dx}\Big|_A$$

<u>Solution</u> :

(1) Calculate the reactions :

$$R_2 = 0 .$$

$$R_1 = \frac{P(2L/3)}{L} = \frac{2P}{3} .$$

$$R_2 = \frac{1}{3}P .$$

$$M_B = \frac{2}{3}P\left(\frac{L}{3}\right) = \frac{2PL}{9} .$$

(2) Calculate the reactions in the conjugate beam :

$$R_2 = 0 :$$

$\curvearrowleft)$ $\Sigma M_A = 0 :$

$$R_3(L) - \frac{1}{2}\left(\frac{2PL}{9EI}\right)\left(\frac{L}{3}\right)\left(\frac{2}{3}\right)\left(\frac{L}{3}\right)$$

$$- \frac{1}{2}\left(\frac{2PL}{9EI}\right)\left(\frac{2L}{3}\right)\left(\frac{L}{3} + \frac{1}{3}\frac{2L}{3}\right) = 0$$

$$\frac{L}{3} + \frac{2L}{9}$$

$$\frac{5L}{9}$$

$$\Rightarrow R_3 = \frac{2PL^2}{243EI} + \frac{10PL^2}{243EI}$$

$$= \frac{12PL^2}{243EI}$$

$$= \frac{4PL^2}{81EI} .$$

$+\uparrow \Sigma F_y = 0: \quad R_1 + \dfrac{4PL^2}{81EI} - \dfrac{1}{2}\left(\dfrac{2PL}{9EI}\right)\left(\dfrac{L}{3}\right) - \dfrac{1}{2}\left(\dfrac{2PL}{9EI}\right)\left(\dfrac{2L}{3}\right) = 0$

$$R_1 = -\dfrac{4PL^2}{81EI} + \dfrac{PL^2}{27EI} + \dfrac{2PL^2}{27EI}$$

$$= \dfrac{5PL^2}{81EI}.$$

$$\theta_{A\ (real\ beam)} = V_{A\ (conjugate\ beam)} = \dfrac{5PL^2}{81EI}.$$

Section:

$+\circlearrowleft \Sigma M_B = 0:$

$$-\dfrac{5PL^2}{81EI}\left(\dfrac{L}{3}\right) + \dfrac{1}{2}\left(\dfrac{2PL}{9EI}\right)\left(\dfrac{L}{3}\right)\left(\dfrac{1}{3}\right)\left(\dfrac{L}{3}\right)$$

$$+ M_B = 0$$

$$M_B = \dfrac{5PL^3}{243EI} - \dfrac{PL^3}{243EI} = \dfrac{4PL^3}{243EI}$$

$$\therefore \ \delta_{B\ (real\ beam)} = M_{B\ (conjugate\ beam)} = \dfrac{4PL^3}{243EI}.$$

Example 6.12 :

Determine the deflection at points B and C for the beam shown in the figure.

Solution :

(1) Calculate the reactions:

$R_2 = 0:$

$+\uparrow \Sigma F_y = 0:$

$$R_1 - w\left(\dfrac{L}{2}\right) = 0$$

$$R_1 = \dfrac{1}{2}wL.$$

$+\circlearrowleft \Sigma M_A = 0:$

$$-m_A - w\left(\dfrac{L}{2}\right)\left(\dfrac{L}{2} + \dfrac{L}{4}\right) = 0$$

$$m_A = -\dfrac{3wL^2}{8}.$$

$V_e = \dfrac{1}{2}wL - w\left(\dfrac{L}{2}\right) = 0.$

$M_B = -\dfrac{3wL^2}{8} + \dfrac{1}{2}wL\left(\dfrac{L}{2}\right) = -\dfrac{wL^2}{8}.$

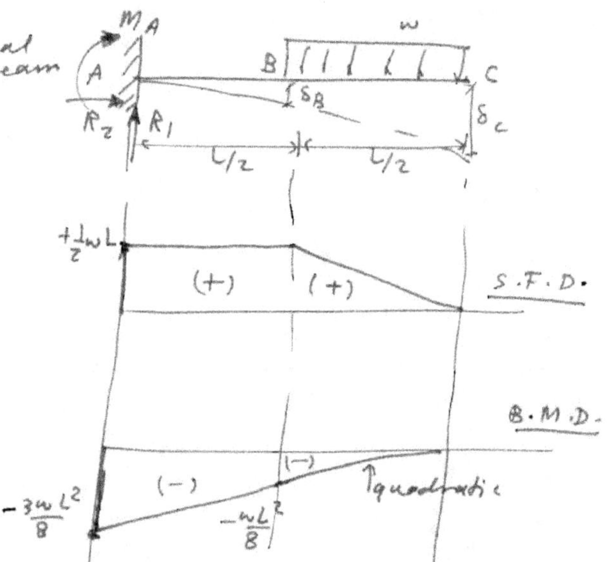

(2) Construct the conjugate beam:

(3) Calculate the reactions in the conjugate beam:

$R_2 = 0$.

$$R_1 = \frac{1}{2}\left(\frac{3wL^2}{8EI} + \frac{wL^2}{8EI}\right)\left(\frac{L}{2}\right)$$

$$+ \frac{1}{3}\left(\frac{wL^2}{8EI}\right)\left(\frac{L}{2}\right)$$

$$= \frac{wL^3}{8EI} + \frac{wL^3}{48EI}$$

$$= \frac{7wL^3}{48EI} \ .$$

$+\circlearrowleft \ \Sigma M_c = 0: \quad M_c - \left(\frac{wL^2}{8EI}\right)\left(\frac{L}{2}\right)\left(\frac{L}{2} + \frac{L}{4}\right) - \frac{1}{2}\left(\frac{wL^2}{4EI}\right)\left(\frac{L}{2}\right)\left(\frac{L}{2} + \frac{2}{3}\left(\frac{L}{2}\right)\right)$

$$- \frac{1}{3}\left(\frac{wL^2}{8EI}\right)\left(\frac{L}{2}\right)\left(\frac{3}{4}\right)\left(\frac{L}{2}\right) = 0$$

$$M_c = \frac{3wL^4}{64\,EI} + \frac{5wL^4}{96\,EI} + \frac{3wL^4}{384\,EI}$$

$$= \left(\frac{18 + 20 + 3}{384\,EI}\right)wL^4$$

$$= \frac{41wL^4}{384\,EI} \ .$$

$$\delta_{C_{(real\ beam)}} = M_{C_{(conjugate\ beam)}} = \frac{41wL^4}{384\,EI} \ .$$

Section:

$+\circlearrowleft \ \Sigma M_B = 0:$

$$M_B - \left(\frac{wL^2}{8EI}\right)\left(\frac{L}{2}\right)\left(\frac{L}{4}\right)$$

$$- \frac{1}{2}\left(\frac{wL^2}{4EI}\right)\left(\frac{L}{2}\right)\left(\frac{2}{3}\right)\left(\frac{L}{2}\right) = 0$$

$$M_B = \frac{wL^4}{64EI} + \frac{wL^4}{48EI} = \frac{7wL^4}{192EI} \ .$$

$$\therefore \ \delta_{B_{(real\ beam)}} = M_{B_{(conjugate\ beam)}} = \frac{7wL^4}{192EI} \ .$$

Example 6.13 :

Calculate the deflection at C for the beam shown in the figure.

Solution : $E = 29 \times 10^3 \ ksi$, $I = 120 \ in^4$.

(1) Calculate the reactions :

$R_2 = 0$:

$+\circlearrowleft \ \Sigma M_A = 0$:

$R_3(20) - 4(30) = 0$

$R_3 = 6 \ k \uparrow$.

$+\uparrow \ \Sigma F_y = 0$:

$R_1 + 6 - 4 = 0$

$\Rightarrow R_1 = -2 \ k$

$R_1 = 2 \ k \downarrow$.

$M_B = -2(20) = -40 \ k\text{-}ft$.

$M_C = -40 + (4)(10) = 0$.

(2) Calculate the reactions in the conjugate beam.

$R_3 = 0$.

Section :

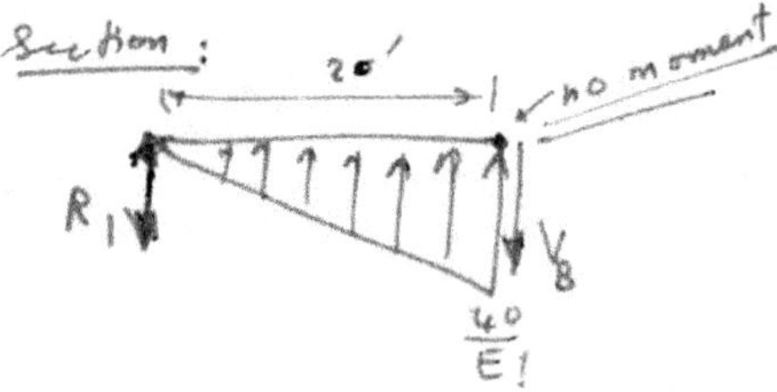

$+\circlearrowright \ \Sigma M_B = 0$:

$+R_1(20) - \frac{1}{2}\left(\frac{40}{EI}\right)(20)\left(\frac{1}{3}\right)(20) = 0$

$R_1 = \frac{133.3}{EI}$.

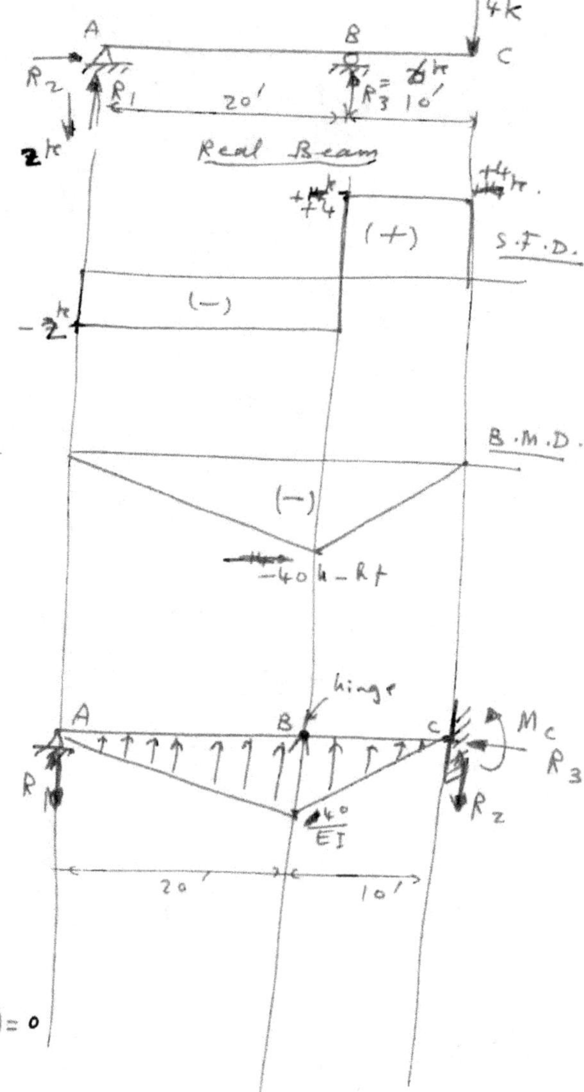

Take the entire beam:

$+\circlearrowright \ \Sigma M_C = 0$:

$\frac{133.3}{EI}(30) - \frac{1}{2}\left(\frac{40}{EI}\right)(20)\left(10 + \frac{20}{3}\right)$

$\quad - \frac{1}{2}\left(\frac{40}{EI}\right)(10)\left(\frac{2}{3}\right)(10) + M_c = 0$

$M_c = \frac{400}{EI}$.

$$E = 29 \times 10^3 \ k/in^2 = 29 \times 10^3 \times (12)^2 \ k/ft^2.$$

$$I = 120 \ in^4 = \frac{120}{(12)^4} \ ft^4.$$

$$\therefore \ \delta_c = M_c = \frac{4001}{EI} = \frac{4001 \ (12)^4}{(29 \times 10^3)(12)^2 (120)} = 0.166 \ ft$$
$$\text{(real beam)} \quad \text{(conjugate beam)}$$
$$= (0.166)(12)$$
$$= 1.99 \ in.$$

Homework 5 $6.8 \ , \ 6.16 \ , \ 6.24 \ , \ 6.34 \ , \ 6.36 \ , \ 6.42 \ , \ 6.44..$
<u>moment-area</u>   <u>conjugate beam</u>

Chapter 7: Deflections : Energy Methods

7.1 Introduction:
* Using the <u>principle of conservation of energy</u>.

7.2 Principle of Conservation of Energy :

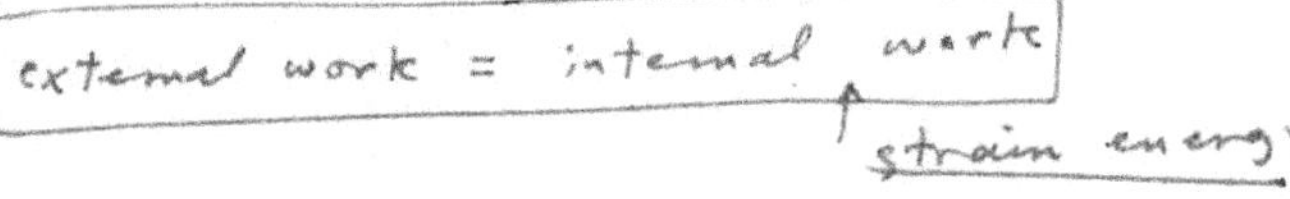

external loads
internal forces } perform work

external work = internal work
                    ↑ strain energy

external work = strain energy

(1) only if the loads are applied <u>gradually</u>,
(2) and if the strains remain <u>elastic</u>.

* <u>Work</u> is defined as the product of force and the displacement, in the direction of the force.

$$\boxed{W = P\Delta}$$

* If the force does <u>not</u> remain constant,

$$W = \int P \, d\Delta$$

$$dW = P \, d\Delta$$

$$W = \int P \, d\Delta$$

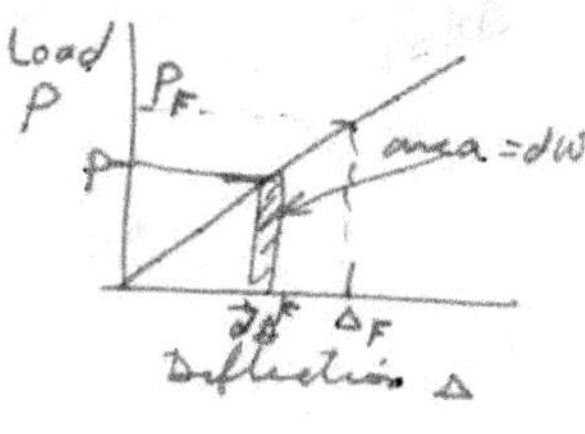

$$E = 29 \times 10^3 \ k/in^2 = 29 \times 10^3 \times (12)^2 \ k/ft^2.$$

$$I = 120 \ in^4 = \frac{120}{(12)^4} \ ft^4.$$

$$\therefore \ \delta_e = M_c = \frac{4001}{EI} = \frac{4001 \ (12)^4}{(29 \times 10^3)(12)^2 \ (120)} = 0.166 \ ft$$
$$\text{(real beam)} \quad \text{(conjugate beam)}$$
$$= (0.166)(12)$$
$$= 1.99 \ in.$$

$\boxed{\text{Homework 5}}$ $\underbrace{6.8 \ , \ 6.16 \ , \ 6.24 \ , \ 6.34}_{\text{moment-area}} \ , \ 6.36 \ , \ \underbrace{6.42 \ , \ 6.44.}_{\text{conjugate beam}}$

Chapter 7: Deflections : Energy Methods

7.1 Introduction :

* Using the principle of conservation of energy.

7.2 Principle of Conservation of Energy :

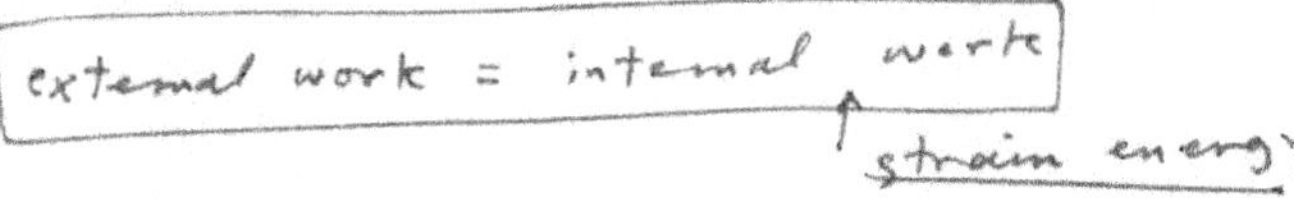

external loads $\Big\}$ perform work
internal forces

$\boxed{\text{external work} = \text{internal work}}$
$$\uparrow$$
$$\underline{\text{strain energy}}$$

$\boxed{\text{external work} = \text{strain energy}}$

(1) only if the loads are applied gradually,
(2) and if the strains remain elastic.

* Work is defined as the product of force and the displacement,
in the direction of the force.

$$\boxed{W = P\Delta}$$

* If the force does not remain constant,

$$W = \int P d\Delta$$

$$dW = P d\Delta$$

$$W = \int P d\Delta$$

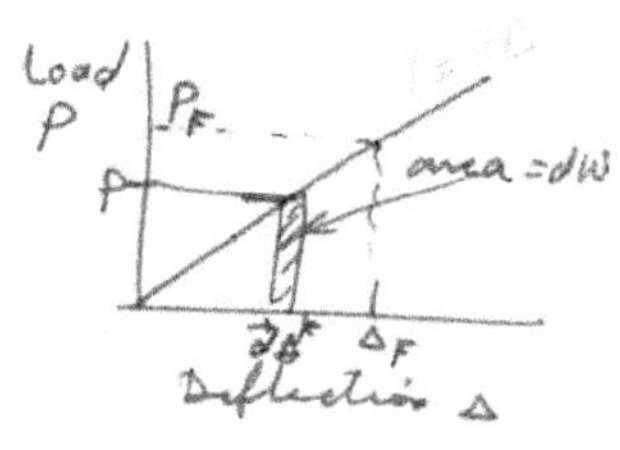

The material obeys Hooke's law, and the load is increased gradually from 0 to P_F :

$$W = \int dW = \int_0^{\Delta_f} P\,d\Delta \quad , \quad \text{but } P = k\Delta \,, \ k \text{ is the slope}$$

$$= \int_0^{\Delta_f} k\Delta\,d\Delta = k \int_0^{\Delta_f} \Delta\,d\Delta = k \left[\frac{\Delta^2}{2}\right]_0^{\Delta_f} = \tfrac{1}{2} k \Delta_f^2$$

but $P_f = k\Delta_f \implies W = \tfrac{1}{2}(k\Delta_f)\Delta_f = \tfrac{1}{2} P_f \Delta_f$.

$$\therefore \quad \boxed{W = \tfrac{1}{2} P_f \Delta_f} \leftarrow \text{one half of the force} \times \text{displacement.}$$

* the work done by a moment M is $\boxed{W = \tfrac{1}{2} M\theta}$, $\theta \equiv$ rotation.

* the work done by a torque T is $\boxed{W = \tfrac{1}{2} T\phi}$, $\phi \equiv$ angle of twist.

7.3 Method of Real Work :

Example :

(1) The load P is gradually applied.

(2) the material of the beam is elastic.

$$\boxed{W_{External} = \tfrac{1}{2} P\delta}$$

Internal work :

* In a truss, only <u>axial forces</u> do internal work.

* In a beam, <u>shears</u> and <u>moments</u> do internal work.

If the span (length) of the beam is <u>three</u> times its depth, the internal work of the shear forces is very small compared with the internal work of the bending moments.

$\therefore$ We use only work done by the bending moment.

$$dW_{Internal} = \tfrac{1}{2} M\,d\theta \ ,$$
$$\text{but} \quad d\theta = \frac{M}{EI}\,dx$$

$$\therefore \quad dW_{Internal} = \frac{1}{2} M \cdot \frac{M}{EI} dx = \frac{M^2}{2EI} dx$$

$$\boxed{\therefore \quad W_{Internal} = \int_0^L \frac{M^2}{2EI} dx}$$

Apply the principle of conservation of energy, $\boxed{\int_0^L \sim = 2\int_0^{L/2}}$

$$W_{External} = W_{Internal}$$

$$\frac{1}{2} P\delta = \int_0^L \frac{M^2}{2EI} dx$$

$W_{Internal} = 2 W^{(1)}_{Internal}$ by symmetry

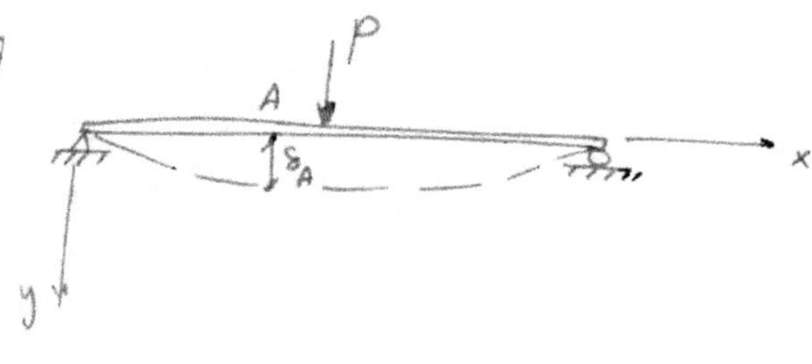

For $\quad 0 < x < L/2, \quad M(x) = \frac{P}{2} x$.   because of symmetry

$$\therefore \quad \frac{1}{2} P\delta = \frac{1}{2EI} (2) \int_0^{L/2} \left(\frac{P}{2} x\right)^2 dx = \frac{1}{8EI} \frac{P^2}{4} \left[\frac{x^3}{3}\right]_0^{L/2}$$

$$\frac{1}{2} P\delta = 2 \left(\frac{P^2}{8EI}\right) \frac{L^3}{3} \cdot \frac{1}{8} \implies \boxed{\delta = \frac{PL^3}{48 EI}}$$

Shortcomings of the Method of Real Work:

(1) It can only be applied to a structure with a single, concentrated load.

(2) It can be used only to calculate the deflection under that load.

7.4 Virtual Work:

* this is the most <u>versatile</u> method for calculating deflections.

* Used for <u>all</u> types of structures: <u>beams, frames & trusses.</u>

<u>Example</u>:

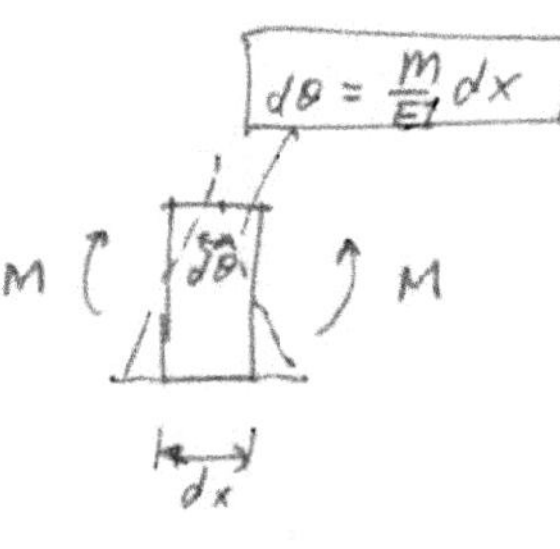

$$d\theta = \frac{M}{EI} dx$$

infinitesimal element

* Consider a __second beam__ with a __unit load__ applied at A.

* In the dummy structure, we use m for moments, and α for rotations.

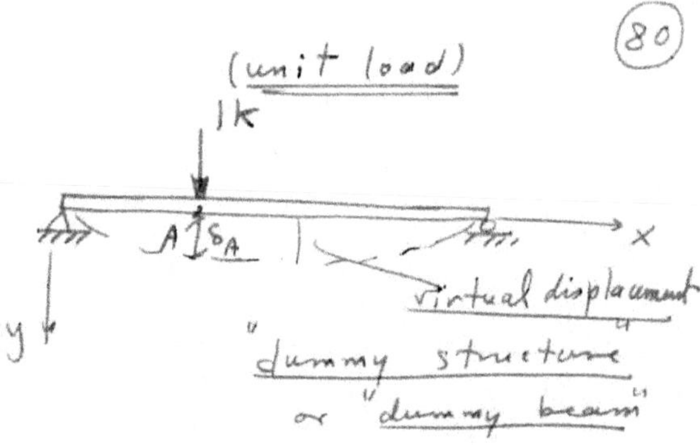

* Apply __virtual displacements__ to the dummy structure.

* Work of the unit load
$$= 1 \cdot \delta_A$$
(The unit load is __constant__ throughout the displacement).

External Virtual Work $= 1 \cdot \delta_A = \delta_A$.

$$dW_{\substack{virtual \\ internal}} = m\, d\theta$$

$$d\theta = \frac{m}{EI}\, dx$$

$$\therefore \text{Internal Virtual Work} = \int_0^L m\, d\theta = \int_0^L m\, \frac{M}{EI}\, dx$$
$$= \int_0^L \frac{Mm}{EI}\, dx$$

* Apply the principle of conservation of energy:
$$\text{External Virtual Work} = \text{Internal Virtual Work}$$
$$\boxed{1 \cdot \delta_A = \int_0^L \frac{Mm}{EI}\, dx}$$

* Virtual work results when one force system acts through displacements caused by another force system.

__Example:__

Calculate the vertical deflection δ_B under the load P?

__Solution:__

$A \xrightarrow{\quad x \quad}$

__Segment AB:__

$0 < x < L/2$

$V = P/2 \;,\qquad v = 0.5$

$M = \dfrac{P}{2}x \;,\qquad m = \dfrac{1}{2}x$

__Segment BC:__

$\xleftarrow{\quad x \quad} C$

$0 < x < L/2$

$V = -P/2 \;,\qquad v = -\dfrac{1}{2}$

$M = \dfrac{P}{2}x \;,\qquad m = \dfrac{1}{2}x$

$$\delta_B = \int_0^L \frac{Mm}{EI}\,dx + \int_0^L \frac{K\,V v}{GA}\,dx$$

use $K = 1.2$ for rectangular cross-sections.

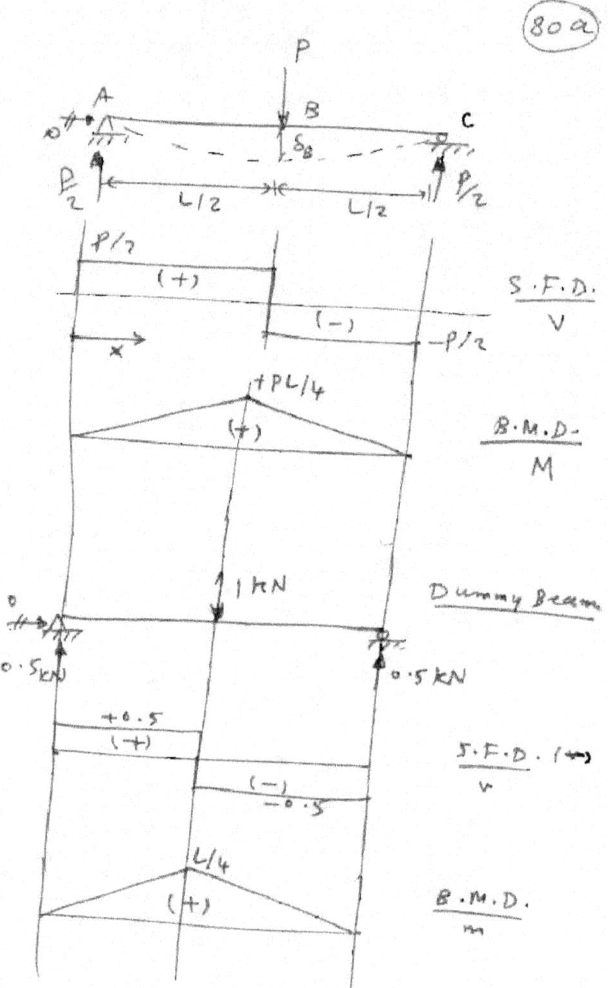

$\therefore$

$$\delta_B = \int_0^{L/2}\frac{\left(\frac{P}{2}x\right)\left(\frac{1}{2}x\right)}{EI}\,dx + \int_0^{L/2}\frac{\left(\frac{P}{2}x\right)\left(\frac{1}{2}x\right)}{EI}\,dx + 1.2\int_0^{L/2}\frac{\left(\frac{P}{2}\right)\left(\frac{1}{2}\right)}{GA}\,dx + 1.2\int_0^{L/2}\frac{\left(-\frac{P}{2}\right)\left(-\frac{1}{2}\right)}{GA}\,dx$$

$$= \frac{P}{4EI}\int_0^{L/2} x^2\,dx + \frac{P}{4EI}\int_0^{L/2} x^2\,dx + \frac{1.2P}{4GA}\int_0^{L/2} dx + \frac{1.2P}{4GA}\int_0^{L/2} dx$$

$$= \frac{P}{2EI}\int_0^{L/2} x^2\,dx + \frac{1.2P}{2GA}\int_0^{L/2} dx$$

$$= \frac{P}{2EI}\left(\frac{1}{3}\left(\frac{L}{2}\right)^3\right) + \frac{1.2P}{2GA}\left(\frac{L}{2}\right) \;=\; \frac{PL^3}{48EI} + \frac{1.2PL}{4GA}$$

For a rectangular cross-section,

$$A = bh$$

$$I = \frac{1}{12} bh^3 .$$

$$\delta_B = \delta_{B\,(Bending)} + \delta_B\,(Shear)$$

$$= \frac{PL^3}{48EI} + \frac{1.2\,PL}{4GA}$$

$$\frac{\delta_{B\,(Shear)}}{\delta_{B\,(Bending)}} = \frac{\dfrac{1.2\,PL}{4GA}}{\dfrac{PL^3}{48EI}} = \frac{1.2(48)EI}{4GAL^2} = \frac{14.4\,EI}{GAL^2}$$

$$G = \frac{E}{2(1+\nu)} \quad , \quad use \ \nu = 0.25$$

$$G = \frac{E}{2(1.25)} = 0.4\,E .$$

$$\therefore \ \frac{\delta_{B\,(Shear)}}{\delta_{B\,(Bending)}} = \frac{14.4\,EI}{(0.4E)AL^2} = \frac{36\,EI}{EAL^2} = \frac{36\,I}{AL^2}$$

$$= \frac{36\left(\frac{1}{12}bd^3\right)}{bd\,L^2} = \frac{3\,d^2}{L^2} = 3\left(\frac{d}{L}\right) . \qquad d = 20\,cm$$

For thin beams $\dfrac{d}{L} = \dfrac{0.20}{5} = 0.04 = \dfrac{4}{100} .$

$$\therefore \ \frac{\delta_{B\,(Shear)}}{\delta_{B\,(Bending)}} = 3\left(\frac{4}{100}\right)^2 = 3\left(\frac{16}{10\,000}\right) = \frac{48}{10\,000} \times 100\% = 4.8\times10^{-3}$$

$$(negligible) \ \approx 0.48\% .$$

Displacement Computation Techniques

* A special technique for calculation of integrals of the type $\int_0^L M\,m\,dx$:

$$\int_0^L M\,m\,dx = A \cdot y_c$$

→ Geometrical evaluation of the integral.

> The product of the multiplication of the two graphs, one of which at least is bounded by a straight line, equals the area bounded by the graph of the arbitrary outline multiplied by the ordinate y_c to the first graph measured along the vertical passing through the centroid of the second one.

This technique is called

Vereshchagin's method.

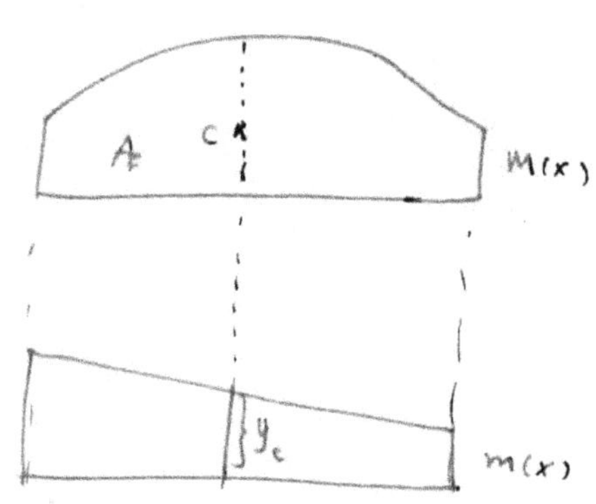

Example 1:

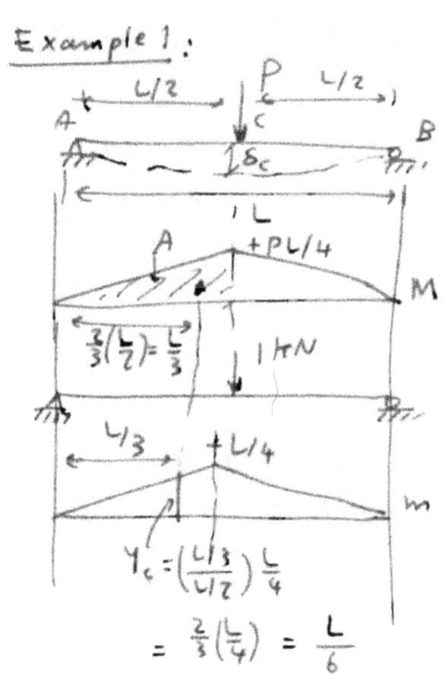

$$\frac{2}{3}\left(\frac{L}{2}\right) = \frac{L}{3}$$

$$y_c = \left(\frac{L/3}{L/2}\right)\frac{L}{4} = \frac{2}{3}\left(\frac{L}{4}\right) = \frac{L}{6}$$

Solution :

$$\delta_c = 2\int_0^{L/2} \frac{M\,m}{EI}\,dx$$

$$= \frac{2}{EI}\int_0^{L/2} M\,m\,dx$$

$$= \frac{2}{EI}\left[A \cdot y_c\right]$$

$$= \frac{2}{EI}\left[\frac{1}{2}\left(\frac{L}{2}\right)\left(\frac{PL}{4}\right)\cdot\frac{L}{6}\right]$$

$$= \frac{PL^3}{48EI}$$

Example 2 :

Calculate the horizontal displacement at the roller (point C) of the portal frame shown in the figure. The moments of inertia of all the members of the frame are indicated in the same figure with E remaining constant throughout.

Solution :

$$A_x = P \leftarrow$$

$+\!\!\curvearrowright \;\; \Sigma M_A = 0 : \quad C_y a - Ph = 0$

$$C_y = Ph/a \cdot \uparrow$$

$$\therefore \; A_y = Ph/a \downarrow$$

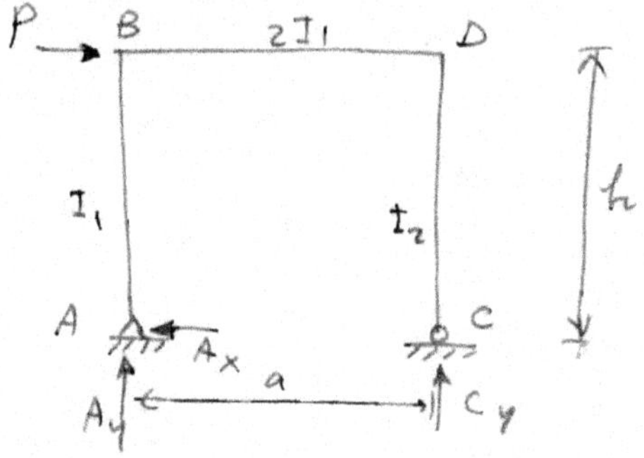

For the dummy frame :

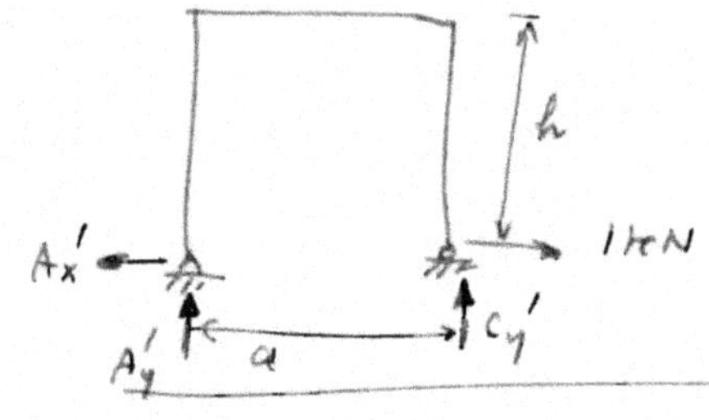

$$A_x' = 1\,hN \leftarrow$$
$$A_y' = C_y' = 0 \cdot$$

Member AB :

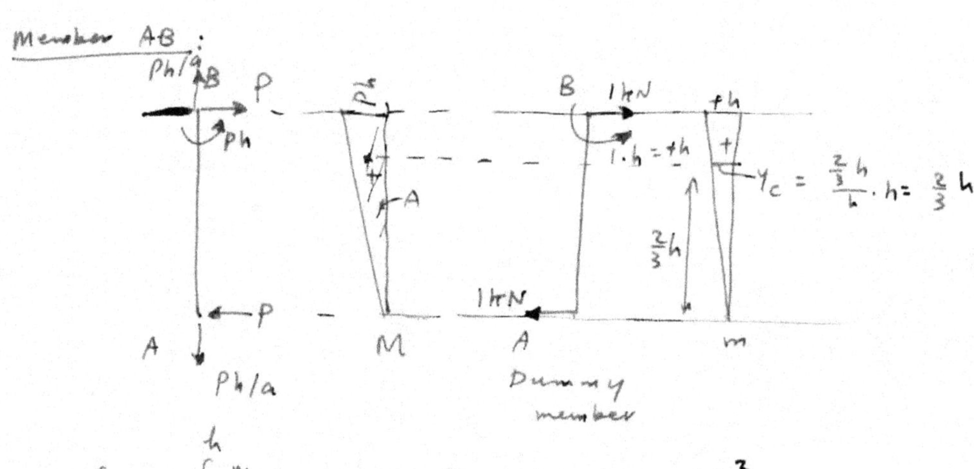

$$\delta_{c_1} = \int_0^h \frac{Mm}{EI_1}\,dx = \frac{1}{EI_1}\left[\frac{1}{2}Ph \cdot h \cdot \frac{2}{3}h\right] = \frac{Ph^3}{3EI_1}$$

Joint B:

$$M_D = Ph - \frac{Ph}{a}(a) = 0.$$

Member BD:

$$\delta_{c_2} = \int_0^a \frac{Mm}{2EI_1}\,dx$$

$$= \frac{1}{2EI_1}\left[\frac{1}{2}\,a \cdot Ph \cdot h\right]$$

$$= \frac{Pah^2}{4EI_1}.$$

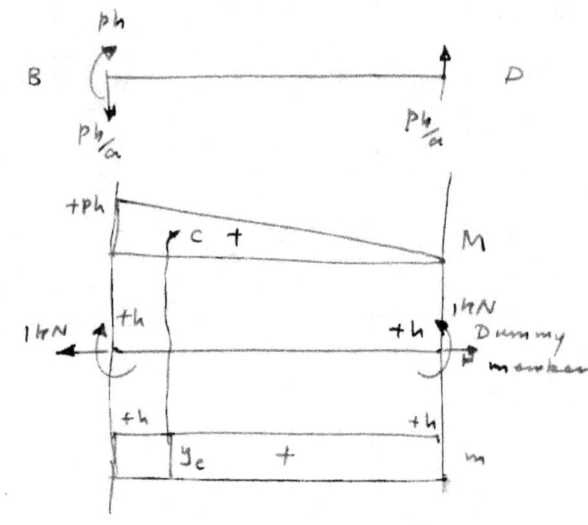

Member DC:

$$\delta_{c_3} = 0.$$

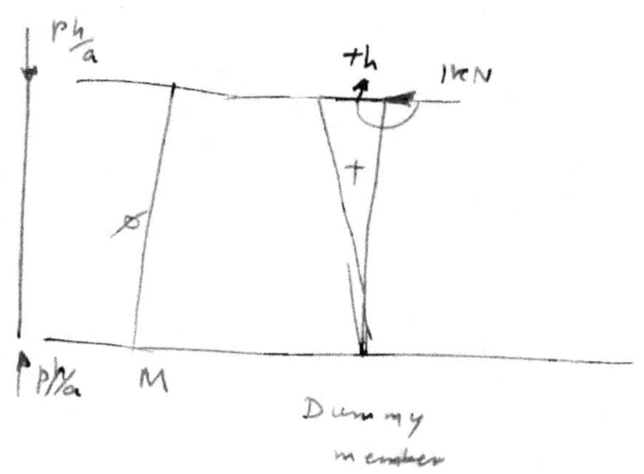

$$\therefore \quad \delta_c = \delta_{c_1} + \delta_{c_2} + \delta_{c_3}$$

$$= \frac{Ph^3}{3EI_1} + \frac{Pah^2}{4EI_1}$$

$$\delta_c = \frac{Ph^2}{EI_1}\left(\frac{h}{3} + \frac{a}{4}\right).$$

Remarks :

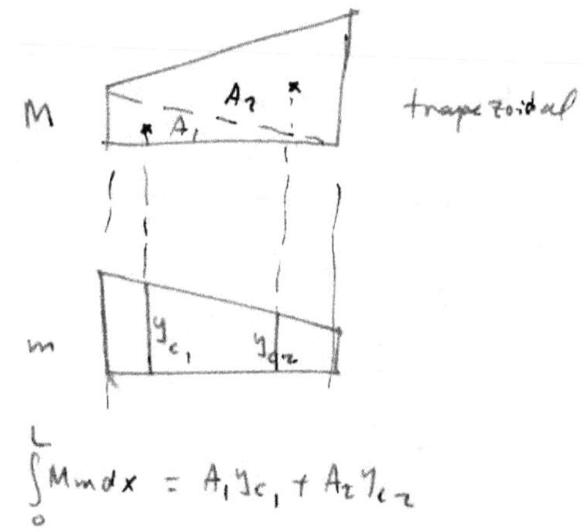

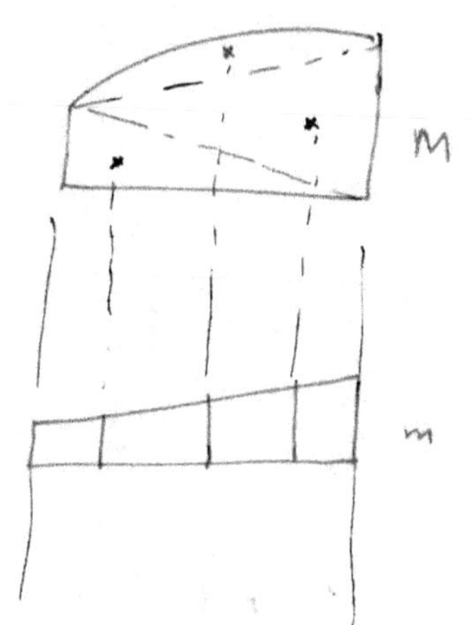

$$\int_0^L M m \, dx = A_1 y_{c_1} + A_2 y_{c_2}$$

Example 3 :

Calculate the deflection δ_c for the beam shown in the figure.

Solution :

It is easier to use integration in this example.

$0 \leq x \leq L/2$,

$$M(x) = -\omega x \cdot \frac{x}{2}$$

$$= -\omega \frac{x^2}{2} .$$

$$m(x) = 0 .$$

$L/2 \leq x \leq L$,

$$M(x) = -\omega x^2 / 2 .$$

$$m(x) = -1 \cdot (x - L/2)$$

$$= -(x - L/2) .$$

$$\therefore \; \delta_c = \int_0^L \frac{M m}{EI} dx = \int_{L/2}^L \frac{\left(-\omega \frac{x^2}{2}\right)(-1)(x - L/2)}{EI} dx$$

$$= \int_{L/2}^L \frac{\omega}{2EI} x^2 (x - L/2) dx$$

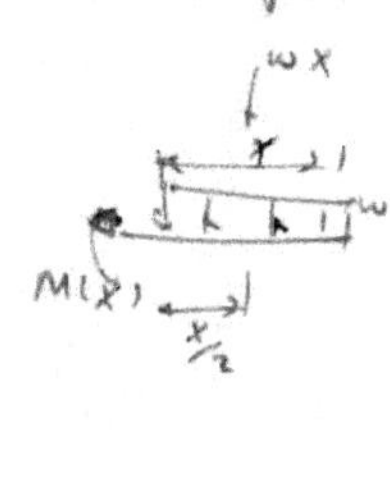

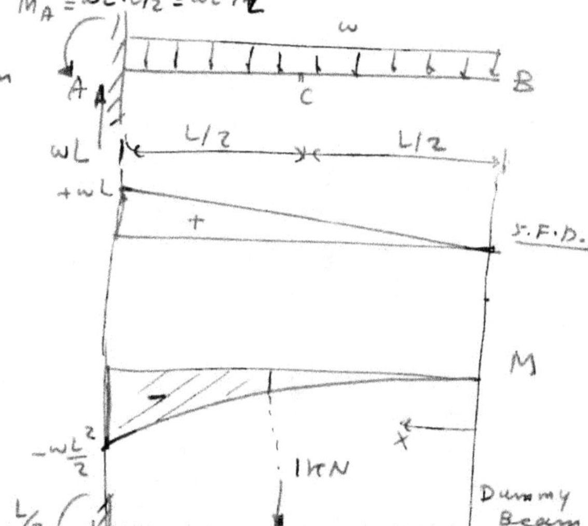

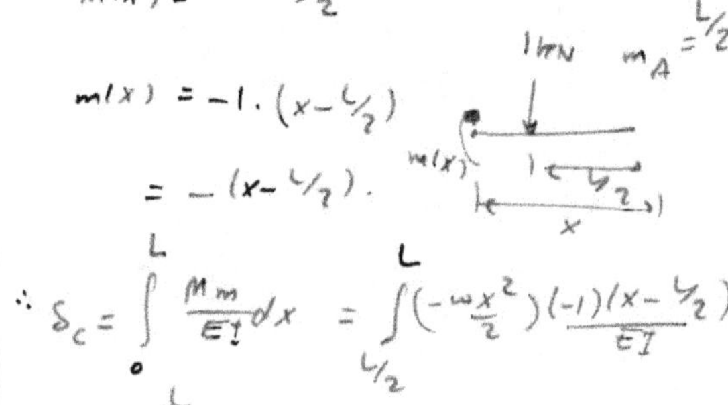

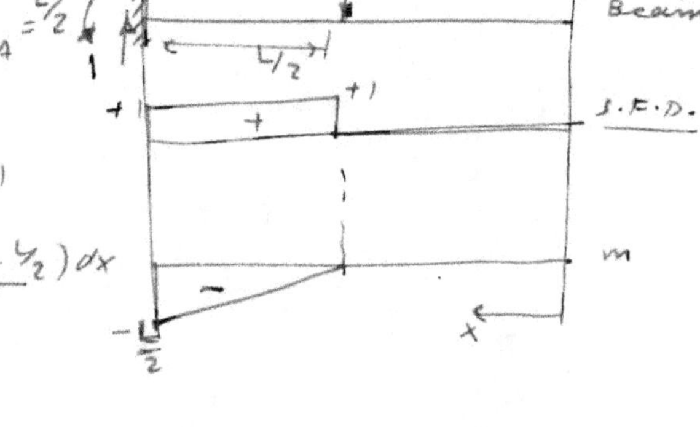

$$= \int_{L/2}^{L} \frac{w}{2EI}\left(x^3 - \frac{L}{2}x^2\right)dx$$

$$= \frac{w}{2EI}\left[\frac{x^4}{4} - \frac{L}{2}\cdot\frac{x^3}{3}\right]_{L/2}^{L} = \frac{w}{2EI}\left[\frac{L^4}{4} - \frac{L}{6}L^3 - \frac{1}{4}\left(\frac{L}{2}\right)^4 + \frac{L}{6}\left(\frac{L}{2}\right)^3\right]$$

$$= \frac{w}{2EI}\left[\frac{1}{4}L^4 - \frac{1}{6}L^4 - \frac{1}{64}L^4 + \frac{1}{48}L^4\right]$$

$$= \frac{w}{2EI}\left(\frac{96L^4 - 64L^4 - 6L^4 + 8L^4}{384}\right)$$

$$\delta_c = \frac{17wL^4}{384EI}\,.$$

Remarks : (Deflections):

① Moment-Area theorems & Conjugate-Beam Method:

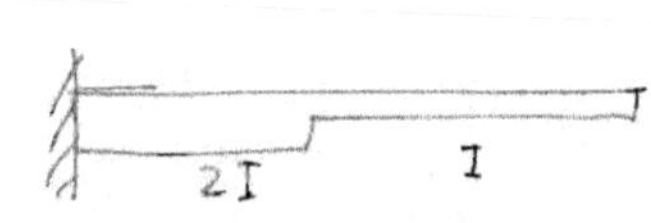

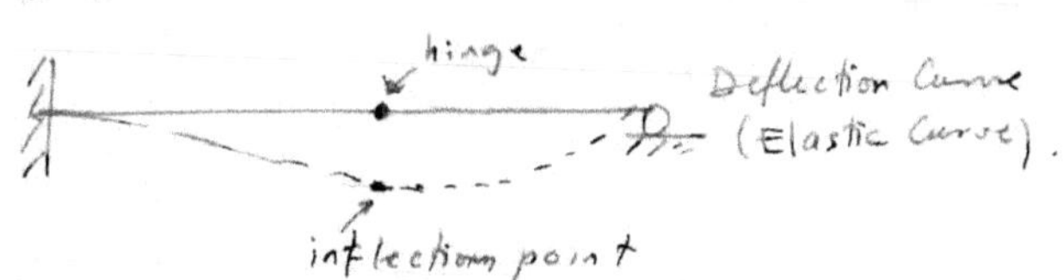

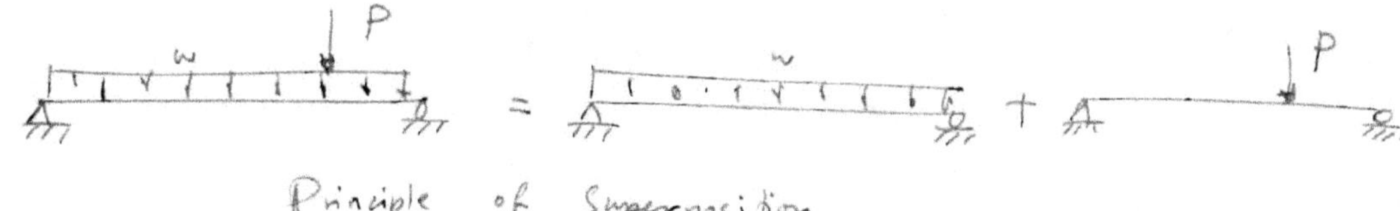

Principle of Superposition

___

② Virtual Work (Unit-Load Method):

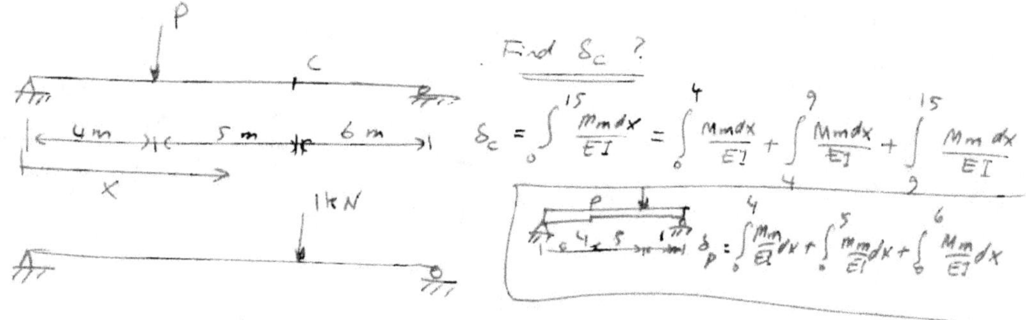

Find δ_C ?

$$\delta_C = \int_0^{15} \frac{Mm\,dx}{EI} = \int_0^4 \frac{Mm\,dx}{EI} + \int_4^9 \frac{Mm\,dx}{EI} + \int_9^{15} \frac{Mm\,dx}{EI}$$

$$\delta_P = \int_0^4 \frac{Mm}{EI}\,dx + \int_0^5 \frac{Mm}{EI}\,dx + \int_0^6 \frac{Mm}{EI}\,dx$$

* Choose the origin x different for each segment:

$$\delta_C = \int_0^4 \frac{Mm\,dx}{EI} + \int_0^5 \frac{Mm\,dx}{EI} + \int_0^6 \frac{Mm\,dx}{EI}$$

___

General Virtual Work Equation:

$$\delta_P = \int_0^L \frac{Mm}{EI}\,dx + \int_0^L \frac{k\,Vv}{GA}\,dx + \int_0^L \frac{Nn}{EA}\,dx$$

neglected for thin, long beams

zero for beams, neglected for frames.

___

Trusses: $M = V = 0$

$$\delta_P = \sum \frac{NnL}{EA}$$

7.5 Application of Virtual Work to Beams and Frames: (81)

Method of Virtual Work : (1) Construct a dummy structure.
(Unit Load Method): (2) Apply a unit load (1k or 1kN)
at the point and in the direction
of the required deflection.

Example 7.1 :

Determine the vertical deflection at the free end of the
cantilever beam shown in the figure.

Solution:

(1) Calculate the reactions and
draw the bending moment
diagram M for the real
beam.

$$M(x) = \begin{cases} -\frac{PL}{2} + Px, & 0 < x < L/2 \\ 0, & 0 < x < L \end{cases}$$

(2) Construct the dummy beam
with a unit load at A.
Calculate the reactions and
draw the bending moment
diagram m for the dummy
beam.

$$m(x) = -L + x$$

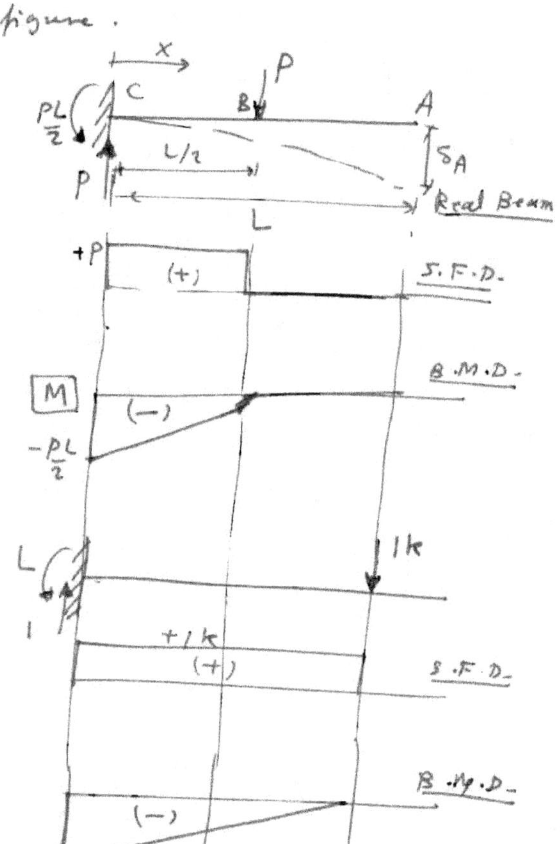

Segment	x = 0	M	m
AB	C	0	$-L + x$
BC	C	$-\frac{PL}{2} + Px$	$-L + x$

$$\delta_A = \int_0^L \frac{Mm}{EI}\,dx = \int_{L/2}^L \frac{Mm}{EI}\,dx + \int_0^{L/2} \frac{Mm}{EI}\,dx$$

$$= \int_0^{L/2} \frac{(0)(x-L)}{EI}\,dx + \int_0^{L/2} \frac{(Px - P\frac{L}{2})(x-L)}{EI}\,dx$$

$$= 0 + \frac{P}{EI} \int_0^{L/2} (x - L/2)(x-L)\,dx$$

$$= \frac{P}{EI} \int_0^{L/2} \left(x^2 - xL - \frac{x}{2}L + \frac{L^2}{2}\right)dx$$

$$= \frac{P}{EI} \int_0^{L/2} \left(x^2 - \frac{3}{2}Lx + \frac{L^2}{2}\right)dx$$

$$= \frac{P}{EI} \left[\frac{x^3}{3} - \frac{3}{2}L\frac{x^2}{2} + \frac{L^2}{2}x\right]_0^{L/2}$$

$$= \frac{P}{EI} \left[\frac{1}{3}\left(\frac{L}{2}\right)^3 - \frac{3}{4}L\left(\frac{L}{2}\right)^2 + \frac{1}{2}L^2\left(\frac{L}{2}\right)\right]$$

$$= \frac{P}{EI} \left[\frac{L^3}{24} - \frac{3L^3}{16} + \frac{L^3}{4}\right]$$

$$= \frac{PL^3}{EI} \left(\frac{2 - 9 + 12}{48}\right)$$

$$= \frac{5PL^3}{48EI}$$

Another way:

Segment	$x=0$	M	m
AB	A	0	x
BC	B	Px	$x + L/2$

$$\delta_A = \int_0^L \frac{Mm}{EI}\,dx = \int_0^{L/2} \frac{(0)(x)}{EI}\,dx + \int_0^{L/2} \frac{Px(x + L/2)}{EI}\,dx$$

$$= 0 + \frac{P}{EI} \int_0^{L/2} \left(x^2 + \frac{1}{2}Lx\right)dx$$

$$= \frac{P}{EI}\left[\frac{x^3}{3} + \frac{1}{2}L\frac{x^2}{2}\right]_0^{L/2} = \frac{P}{EI}\left[\frac{1}{3}\left(\frac{L}{2}\right)^3 + \frac{1}{4}L\left(\frac{L}{2}\right)^2\right]$$

$$= \frac{P}{EI}\left(\frac{L^3}{24} + \frac{L^3}{16}\right)$$

$$= \frac{5PL^3}{48EI}$$

Sign Convention :

① The same criterion must be used to define positive bending for both the real and the dummy structure; i.e. Both M and m must be assumed positive when they produce compression on the same side of the member.

② A positive answer for the deflection indicates that the direction of the deflection is the same as that of the dummy load. A negative answer means that the deflection is opposite in direction to the dummy load.

Example 7.2 :

Determine the vertical deflection at C for the beam shown in the figure. $E = 29 \times 10^3$ ksi, $I = 100$ in^4.

Solution :

(1) Calculate the reactions and find an expression for M.

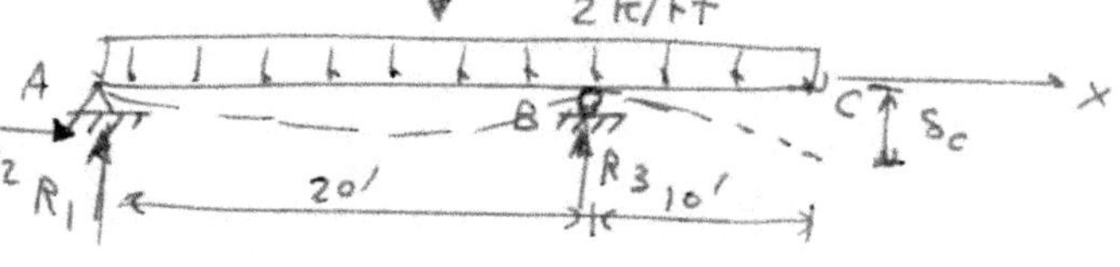

$R_2 = 0$.

Take the origin of coordinates at A, $(x=0)$.

$+\curvearrowleft \ \Sigma M_A = 0$:

$$R_3 (20) - 60(15) = 0$$
$$R_3 = 45 \ k \uparrow .$$
$$R_1 = 15 \ k \uparrow$$

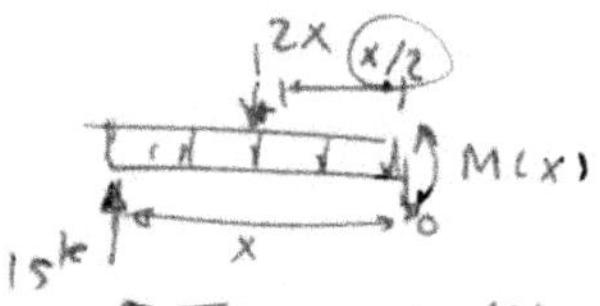

$+\curvearrowleft \ \Sigma M_o = 0$: $M(x) - 15x + 2x\left(\dfrac{x}{2}\right) = 0$

$$M(x) = 15x - x^2 \ , \quad 0 < x < 20$$

$+\curvearrowleft \ \Sigma m_x = 0$:

$$- m(x) - 2x\left(\dfrac{x}{2}\right) = 0$$
$$m(x) = -x^2 \ , \quad 0 \leq x \leq 10$$

(2) Construct the dummy beam and find an expression for m.

$R_2 = 0$.

$+\circlearrowright \Sigma M_A = 0:$

$\qquad R_3 (20) - 1(30) = 0$

$\qquad R_3 = 1.5 \text{ k} \uparrow$

$+\uparrow \Sigma F_y = 0:$

$\qquad R_1 + 1.5 - 1 = 0$

$\qquad R_1 = -0.5 \text{ k}$

$\qquad \underline{R_1 = 0.5 \text{ k} \downarrow}$

$+\circlearrowright \Sigma M_0 = 0:$

$\qquad M(x) + 0.5(x) = 0$

$\qquad M(x) = -\tfrac{1}{2} x, \quad 0 < x < 20$

$+\circlearrowright \Sigma M_x = 0:$

$\qquad -M(x) - 1 x = 0$

$\qquad M(x) = -x, \quad 0 < x < 10$

$$\therefore \; S_c = \int_0^L \frac{Mm}{EI} \, dx = \int_0^{20} \frac{(15x - x^2)(-\tfrac{1}{2}x)}{EI} \, dx + \int_0^{10} \frac{(-x^2)(-x)}{EI} \, dx$$

$$= -\frac{1}{2EI} \int_0^{20} (15x^2 - x^3) \, dx + \frac{1}{EI} \int_0^{10} x^3 \, dx$$

$$= -\frac{1}{2EI} \left[15 \frac{x^3}{3} - \frac{x^4}{4} \right]_0^{20} + \frac{1}{EI} \left[\frac{1}{4} x^4 \right]_0^{10}$$

$$= -\frac{1}{2EI} \left[\frac{15}{3}(20)^3 - \frac{1}{4}(20)^4 \right] + \frac{1}{4EI} (10)^4$$

$$= 0 + \frac{(10)^4}{4EI}$$

$$= \frac{2500}{EI}$$

$$\therefore \; \underline{S_c = \frac{2500 \times (12)^2}{29 \times 10^3 \times 100} = 0.124 \text{ ft} = 1.49 \text{ in.}}$$

Example 7.3 :

Determine both the vertical and the horizontal deflections at A for the frame shown in the figure.

Solution : $E = 200 \times 10^6 \ kN/m^2$.
$I = 200 \times 10^6 \ mm^4$.

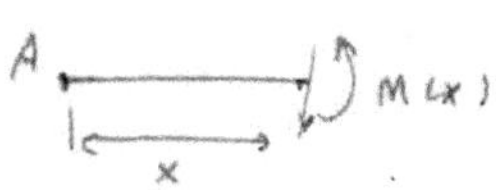

(1) Calculate the reactions and find an expression for M.

$R_2 = 0$.
$R_1 = 50 \ kN \uparrow$.
$M_D = 50(2) = 100 \ kN \cdot m$.

Segment AB :

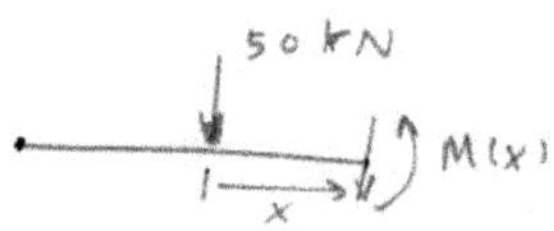

$M(x) = 0$, $0 < x < 2$

Segment BC :

$M(x) = -50x$, $0 < x < 2$

Segment CD :

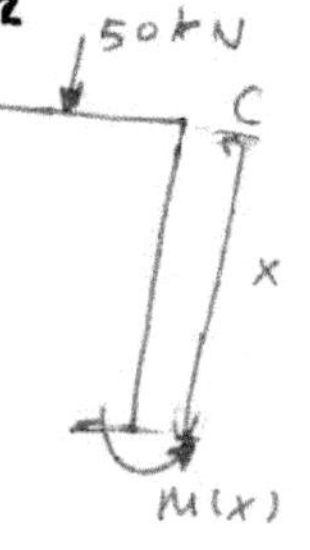

$M(x) = -100$, $0 < x < 5$

---

(2) Construct the first dummy frame :

Segment AB :

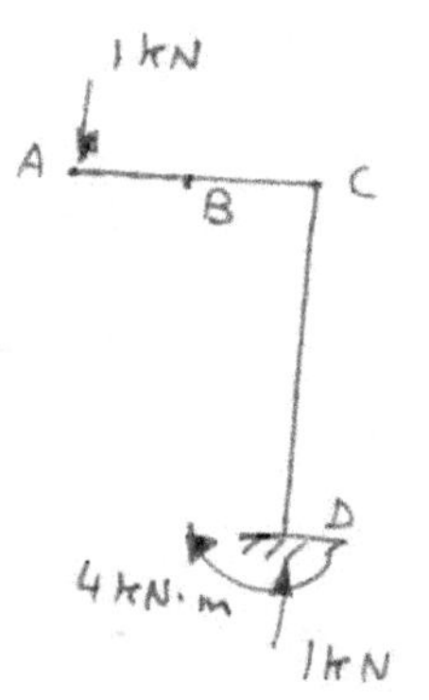

$m(x) = -x$, $0 < x < 2$

Segment BC :

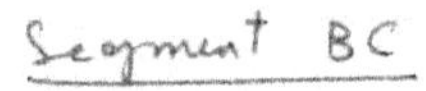

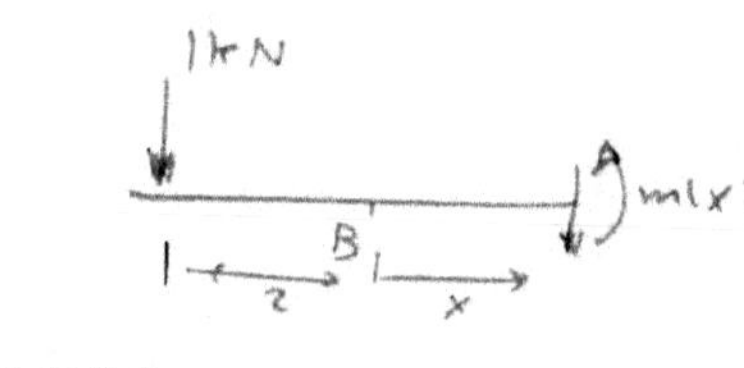

$m(x) = -1(x+2)$

$\quad = -(2+x), \quad 0 < x < 2$

Segment CD :

$m(x) = -1(4) = -4,$

$\quad\quad 0 < x < 5.$

$$\delta_{A_V} = \int_0^L \frac{Mm}{EI}\,dx = \int_0^2 \frac{(0)(-x)}{EI}\,dx + \int_0^2 \frac{(-50x)[-(2+x)]}{EI}\,dx + \int_0^5 \frac{(-100)(-4)}{EI}\,dx$$

$$= 0 + \frac{50}{EI}\int_0^2 (2x+x^2)\,dx + \frac{400}{EI}\int_0^5 dx$$

$$= \frac{50}{EI}\left[x^2 + \frac{x^3}{3}\right]_0^2 + \frac{400}{EI}\left[x\right]_0^5$$

$$= \frac{50}{EI}\left[4+\frac{8}{3}\right] + \frac{400}{EI}[5]$$

$$= \frac{2333}{EI}$$

$$= \frac{2333}{(200\times10^6)(200\times10^6)\times(10^{-3})^4} = 0.058\,m = 5.8\,cm.$$

(3) Construct the second dummy structure (frame):

Segment AB :

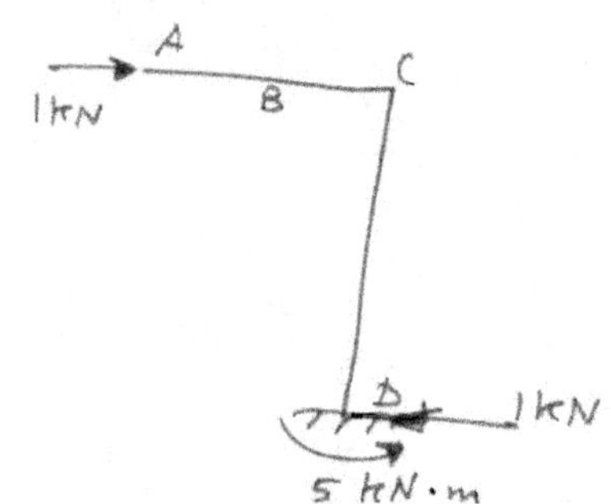

$m(x) = 0, \quad\quad 0 < x < 2$

Segment BC :

$m(x) = 0, \quad 0 < x < 2$

Segment CD :

$$m(x) = 1 \cdot x = x, \qquad 0 < x < 5$$

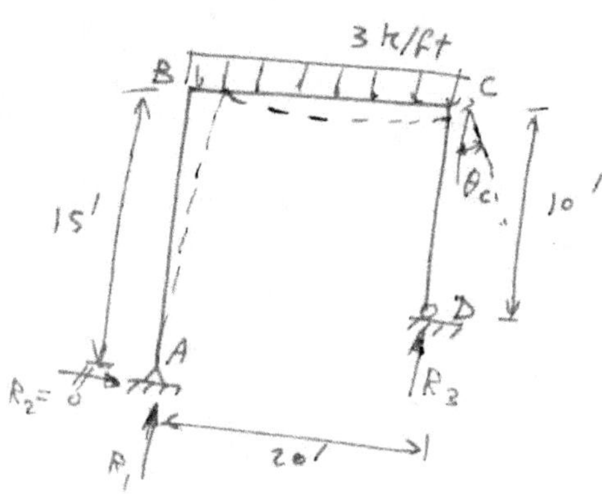

$$\delta_{A_H} = \int_0^L \frac{Mm}{EI}\, dx$$

$$= 0 + 0 + \int_0^5 \frac{(-100)(x)}{EI}\, dx$$

$$= -\frac{100}{EI}\left[\frac{x^2}{2}\right]_0^5 \quad = \quad -\frac{100}{EI}\left(\frac{(5)^2}{2}\right) = \frac{-1250}{EI}.$$

$$\therefore\ \delta_{A_H} = \frac{-1250}{(200\times10^6)(200\times10^6)(10^{-3})^4} = -0.03125\ m = -3.125\ cm.$$

(the horizontal deflection is to the left).

Example 7.4 :

Determine the rotation of joint C for the frame shown in the figure.

$$E = 29\times10^3\ ksi$$
$$I = 240\ in^4.$$

Solution :

(1) $R_2 = 0$.

$$R_1 = R_3 = \frac{3(20)}{2} = 30k \uparrow.$$

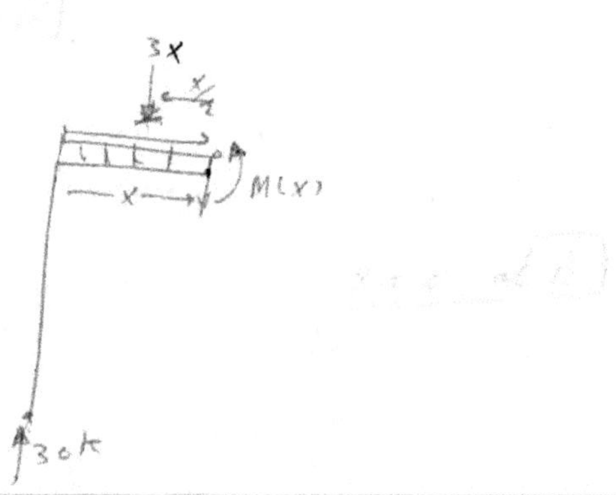

Segment AB :

$$M(x) = 0,$$
$$0 < x < 15$$

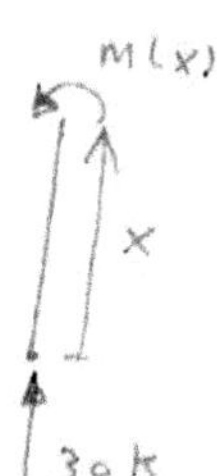

Segment BC :

$$M(x) = 30x - 3x\left(\frac{x}{2}\right)$$
$$= 30x - \frac{3}{2}x^2, \qquad 0 < x < 20$$

<u>Segment CD</u> :

$M(x) = 0, \quad 0 < x < 10$

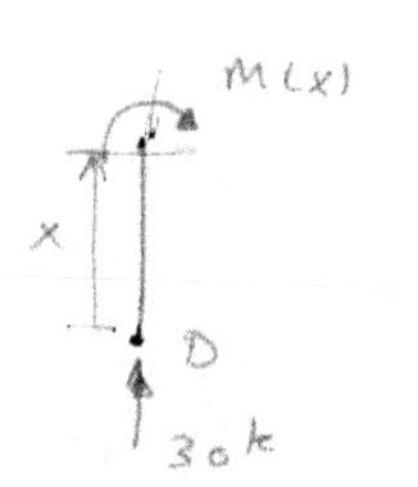

(2) Construct the dummy frame and find an expression for m.

$R_2 = 0$.

$+\circlearrowleft \; \Sigma M_A = 0:$

$R_3(20) - 1 = 0$

$R_3 = \dfrac{1}{20} = 0.05 \; k \uparrow$

$\therefore \; R_1 = 0.05 \; k \downarrow$

<u>Segment AB</u> :

$M(x) = 0, \quad 0 < x < 15$

<u>Segment BC</u> :

$M(x) = -0.05x,$

$\quad 0 < x < 20$

<u>Segment CD:</u>

$M(x) = 0, \quad 0 < x < 10$

$\theta_c = \displaystyle\int_0^L \frac{Mm}{EI}\,dx = 0 + \frac{1}{EI}\int_0^{20} \left(30x - \frac{3}{2}x^2\right)(-0.05x)\,dx + 0$

$\quad = \dfrac{-0.05}{EI}\displaystyle\int_0^{20}\left(30x^2 - \frac{3}{2}x^3\right)dx$

$\quad = \dfrac{-0.05}{EI}\left[10x^3 - \frac{3}{8}x^4\right]_0^{20} = \dfrac{-0.05}{EI}\left[10(20)^3 - \frac{3}{8}(20)^4\right]$

$\quad = \dfrac{-1000}{EI}$

$\quad = \dfrac{-1000 \times (12)^2}{29,000 \times 240} = \underline{-0.021 \; rad.}$

the rotation is counterclockwise.

Example 7.5:

Determine the vertical deflection at A for the structure shown in the figure.

Solution:

$E = 30 \times 10^3$ ksi

$G = 12 \times 10^3$ ksi

$I = 144$ in^4.

$J = 288$ in^4.

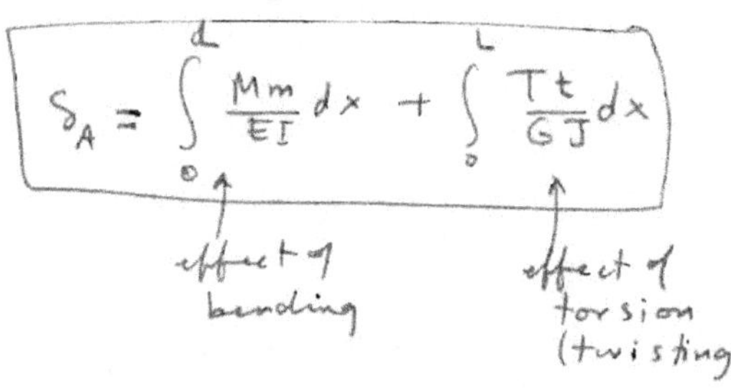

$$\delta_A = \int_0^d \frac{Mm}{EI}\,dx + \int_0^L \frac{Tt}{GJ}\,dx$$

effect of bending

effect of torsion (twisting)

* there is <u>no</u> need to find the reactions in this problem.

* Internal work is due to both <u>bending</u> and <u>torsion</u>.

* Construct the dummy structure.

Segment AB:

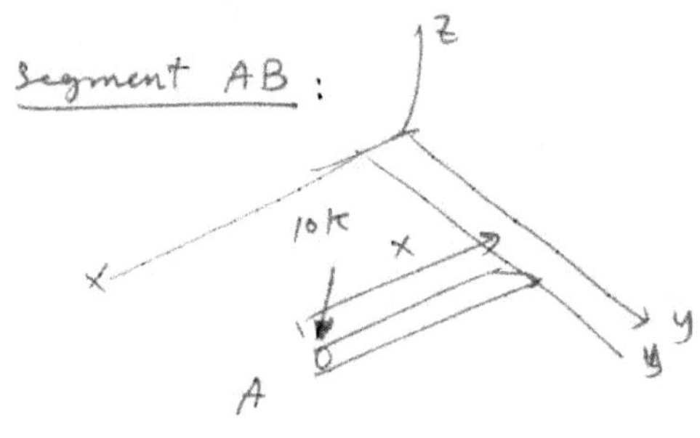

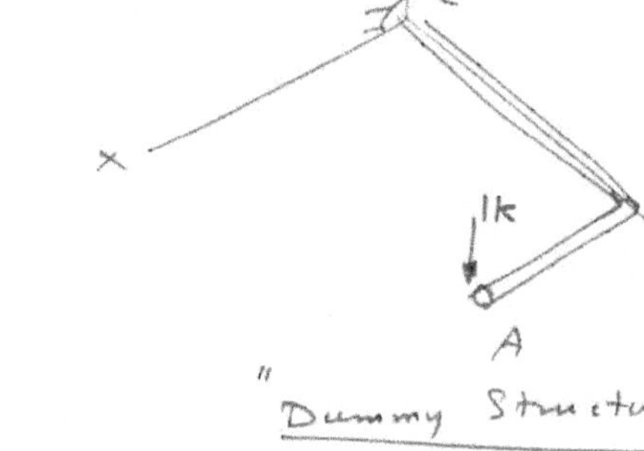

$M_y(x) = 10x$, $0 < x < 5$

$m_y(x) = 1 \cdot x = x$, $0 < x < 5$.

Segment BC:

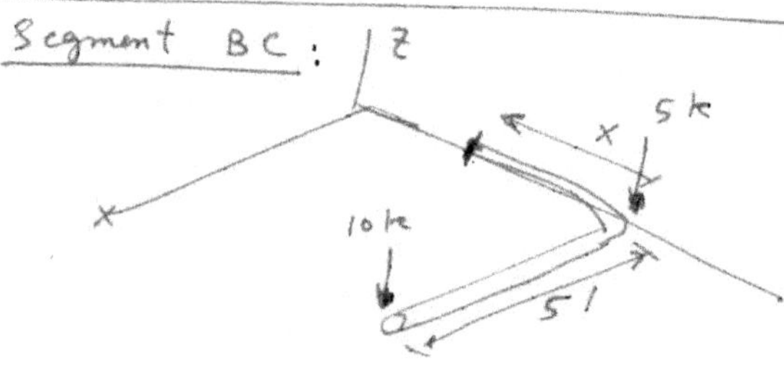

$M_x(x) = -10x - 5x$ $0 < x < 5$
$ = -15x$

$T_y(x) = 10(5) = 50$ k-ft, $0 < x < 5$.

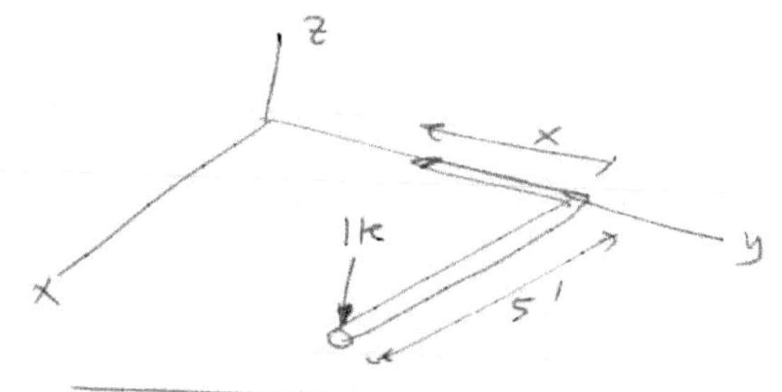

$$m_x(x) = -1 \cdot x = -x, \qquad 0 < x < 5$$

$$t_y(x) = (1)(5) = 5, \qquad 0 < x < 5 .$$

$$\delta_A = \int_0^L \frac{Mm}{EI}\,dx \;+\; \int_0^L \frac{Tt}{GJ}\,dx$$

$$= \int_0^5 \frac{(10x)(x)}{EI}\,dx \;+\; \int_0^5 \frac{(-15x)(-x)}{EI}\,dx \;+\; \int_0^5 \frac{(50)(5)}{GJ}\,dx$$

$$= \frac{10}{EI}\int_0^5 x^2\,dx \;+\; \frac{15}{EI}\int_0^5 x^2\,dx \;+\; \frac{250}{GJ}\int_0^5 dx$$

$$= \frac{10}{EI}\left[\frac{(5)^3}{3}\right] + \frac{15}{EI}\left[\frac{(5)^3}{3}\right] + \frac{250}{GJ}(5)$$

$$= \frac{1042}{EI} \;+\; \frac{1250}{GJ}$$

$$= \frac{1042 \times (12)^2}{(30\times10^3)\times 144} \;+\; \frac{1250 \times (12)^2}{(12\times10^3)\times 288}$$

$$= 0.0347 \;+\; 0.0521 \qquad \uparrow \text{ significant effect}$$

$$= 0.0868 \text{ ft}$$

$$= 1.04 \text{ in} .$$

7.6 Deflection of Trusses Using Virtual Work:

S : Member forces in the real truss

u : member forces in the dummy truss.

L : length of each member.

A : Cross-sectional area of each member

E : modulus of elasticity .

Axial Deformation in a member $= \dfrac{SL}{EA}$

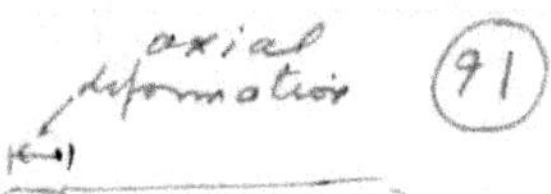

$$\Delta = \sum \frac{u\,S\,L}{EA}$$

<u>Example 7.6</u> :

Determine the vertical deflection of point g for the truss shown in the figure.

<u>Solution</u> :

$E = 30 \times 10^{3}$ ksi.

the truss is <u>symmetric</u>.

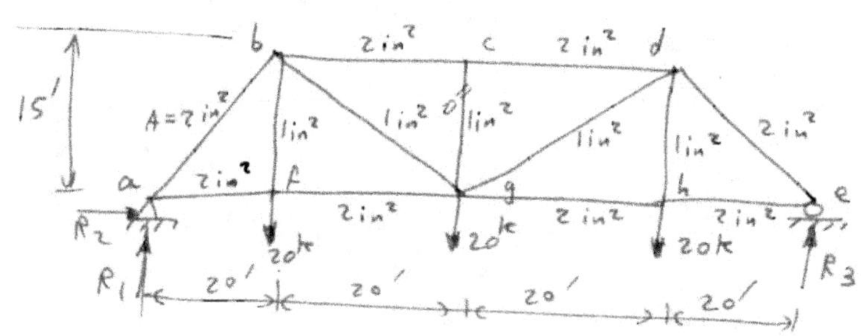

(1) Find the member forces in the real truss.

First find the reactions:

From <u>symmetry</u>, $R_2 = 0$, $R_1 = R_3 = \dfrac{20+20+20}{2} = 30\,k$.

<u>Use the method of joints</u>:

<u>Joint a</u> :

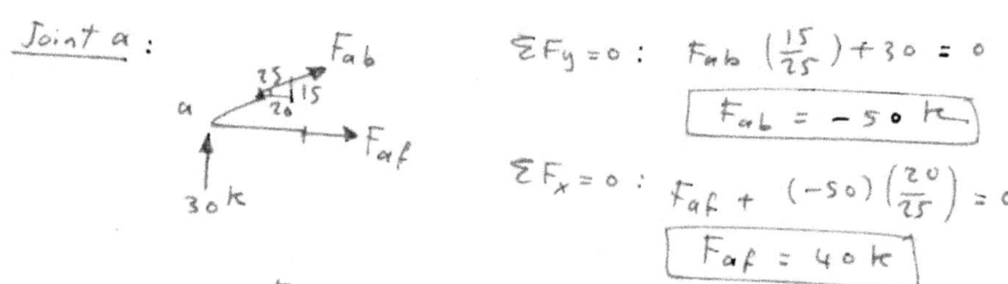

$\Sigma F_y = 0:\quad F_{ab}\left(\dfrac{15}{25}\right) + 30 = 0$

$$\boxed{F_{ab} = -50\ k}$$

$\Sigma F_x = 0:\quad F_{af} + (-50)\left(\dfrac{20}{25}\right) = 0$

$$\boxed{F_{af} = 40\ k}$$

<u>Joint f</u> :

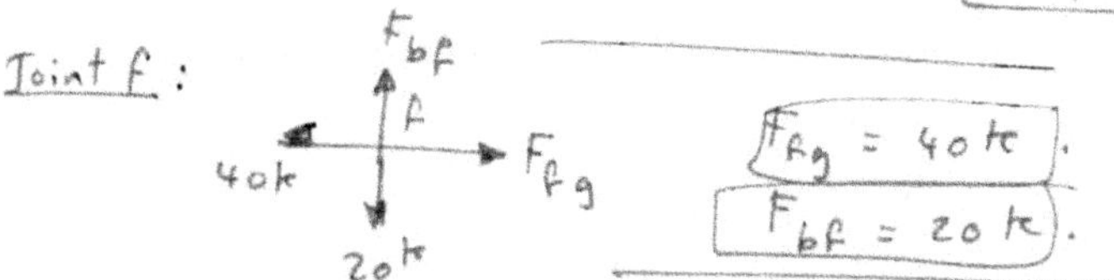

$$\boxed{F_{fg} = 40\ k}$$

$$\boxed{F_{bf} = 20\ k}$$

<u>Joint b</u> :

$\Sigma F_y = 0:$

$50\left(\dfrac{15}{25}\right) - 20 - F_{bg}\left(\dfrac{15}{25}\right) = 0$

$$\boxed{F_{bg} = 16.7}$$

$$\Sigma F_x = 0: \quad F_{bc} + 16.7\left(\frac{20}{25}\right) + 50\left(\frac{20}{25}\right) = 0$$

$$\boxed{F_{bc} = -53.4 \ k}$$

Joint C :

$$\boxed{F_{cg} = 0}$$

Construct the dummy truss:

$$R_2 = 0.$$

$$R_1 = R_3 = \frac{1}{2} = 0.5 \ k.$$

Calculate all the member forces:

Joint a :

$$\Sigma F_y = 0: \quad F_{ab}\left(\frac{15}{25}\right) + 0.5 = 0$$

$$\boxed{F_{ab} = -0.83 \ k.}$$

$$\Sigma F_x = 0: \quad F_{af} + (-0.83)\left(\frac{20}{25}\right) = 0$$

$$\boxed{F_{af} = 0.66 \ k.}$$

Joint f :

$$\boxed{F_{bf} = 0}$$

$$\boxed{F_{fg} = 0.66 \ k}$$

$$\Sigma F_y = 0: \quad 0.83\left(\frac{15}{25}\right) - F_{bg}\left(\frac{15}{25}\right) = 0$$

$$\boxed{F_{bg} = 0.83}$$

Joint b :

$$\Sigma F_x = 0: \quad 0.83\left(\frac{20}{25}\right) + 0.83\left(\frac{20}{25}\right) + F_{bc} = 0$$

$$\boxed{F_{bc} = -1.33}$$

Joint C :

$$\boxed{F_{eg} = 0}$$

Member	A (in²)	L (ft)	S (k)	u (k)	$\dfrac{SuL}{A}$	
ab	2	25	−50	−0.83	518.75	
af	2	20	40	0.66	264.0	
fg	2	20	40	0.66	264.0	
bf	1	15	20	0	0	
bg	1	25	16.7	0.83	346.5	
bc	2	20	−53.4	−1.33	710.2	
cg	1	15	0	0	0	

$$\sum SuL/A = 2103.45$$

because of symmetry

$$\delta_g = 2\sum \frac{SuL}{EA} = \frac{1}{E}\left(2\sum \frac{SuL}{A}\right) = 2\left(\frac{2103.45}{E}\right)$$

$$= \frac{4206.9}{E}$$

$$= \frac{4206.9}{30\times10^3}$$

$$= 0.140 \ ft$$

$$= \underline{1.68 \ in.}$$

7.7 Deflection of Structures Consisting of Flexural Members and Axially-Loaded Members:

$$\delta_c = \int_0^L \frac{Mm}{EI} dx + \sum \frac{SuL}{EA}$$

Example 7.7 :

Determine the vertical deflection of point C for the structure shown in the figure.

Solution :

$$E = 200 \times 10^6 \ kN/m^2$$
$$I_b = 80 \times 10^6 \ mm^4.$$
$$A_c = 150 \ mm^2$$

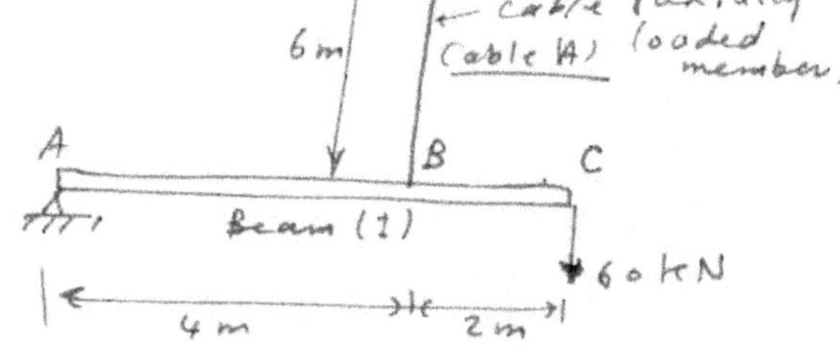

Construct the dummy structure:

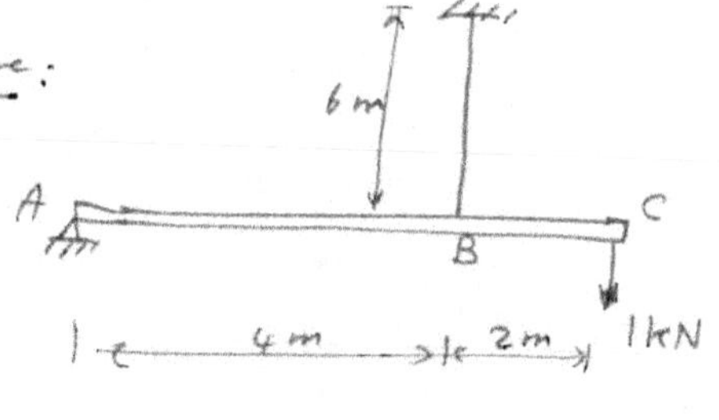

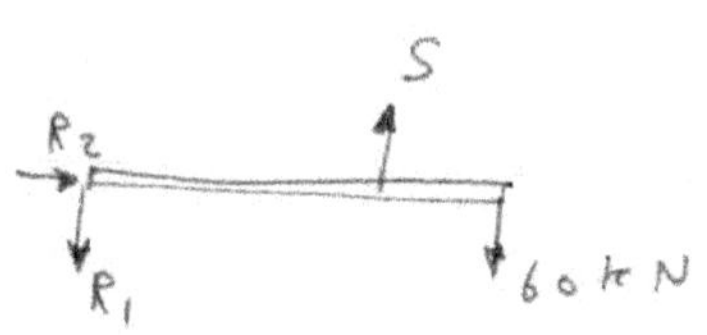

$R_2 = 0:$

$+) \Sigma m_A = 0: \quad S(4) - 60(6) = 0$

$\qquad\qquad S = 90 \text{ kN}.$

$+\uparrow \Sigma F_y = 0: \quad -R_1 - 60 + 90 = 0$

$\qquad\qquad R_1 = 30 \text{ kN} \downarrow.$

$R_2 = 0:$

$+) \Sigma m_A = 0: \quad u(4) - 1(6) = 0$

$\qquad\qquad u = 1.5 \text{ kN}.$

$+\uparrow \Sigma F_y = 0: \quad -R_1 - 1 + 1.5 = 0$

$\qquad\qquad R_1 = 0.5 \text{ kN} \downarrow.$

Segment AB :

$M(x) = -30x, \quad 0 < x < 4$

30 kN

$m(x) = -0.5x, \quad 0 < x < 4$

0.5 kN

Segment BC :

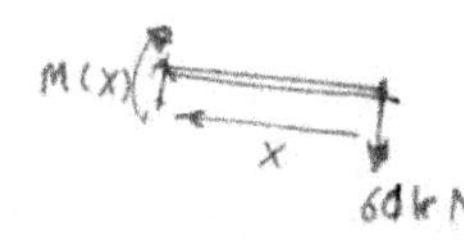

60 kN

$M(x) = -60x, \quad 0 < x < 2$

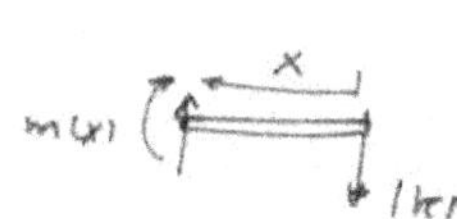

1 kN

$m(x) = -x, \quad 0 < x < 2$

$\therefore \delta_c = \int_0^L \dfrac{Mm}{EI}dx + \Sigma \dfrac{SuL}{EA}$

$= \int_0^4 \dfrac{(-30x)(-0.5x)dx}{EI} + \int_0^2 \dfrac{(-60x)(-x)dx}{EI} + \dfrac{(90)(1.5)(6)}{EA}$

$$= \frac{15}{EI} \int_0^4 x^2 dx + \frac{60}{EI} \int_0^2 x^2 dx + \frac{810}{EA}$$

$$= \frac{15}{EI} \left(\frac{(4)^3}{3} \right) + \frac{60}{EI} \left(\frac{(2)^3}{3} \right) + \frac{810}{EA}$$

$$= \frac{480}{EI} + \frac{810}{EA}$$

$$= \frac{480}{(200 \times 10^6)(80 \times 10^6) \times (10^{-3})^4} + \frac{810}{(200 \times 10^6)(450) \times (10^{-3})^2}$$

$$= 0.03 + 0.027$$

$$= 0.057 \ m$$

$$= 5.7 \ cm.$$

Homework 6	7.10, 7.12, 7.20, 7.28, 7.32, 7.36.

Chapter 8 : Influence Lines

8.1 Introduction :

* To determine the position of the (moving) load that produces the maximum stresses in a structure.

8.2 Influence Lines Defined:

* Influence lines provide us with a systematic procedure for determining how the force in in a given part of a structure varies as the applied load moves about on the structure.

Example :

Determine the value of the reaction R_B for the simply-supported beam shown due to a unit moving load.

$$R_B = \frac{1 \cdot x}{8} = \frac{x}{8} .$$

x	0	2	4	6	8
R_B	0	0.25	0.5	0.75	1.0

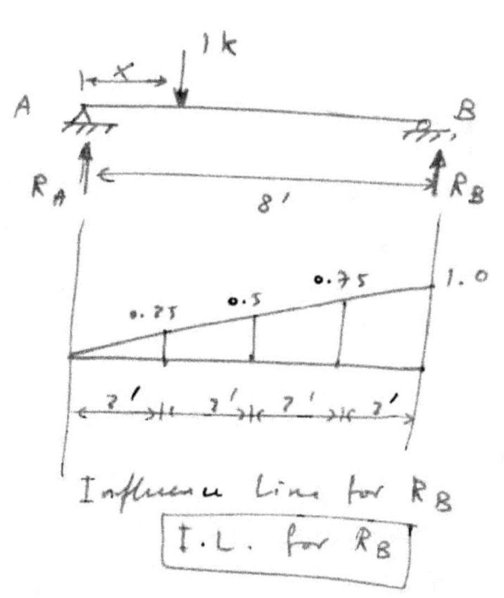

The y value of the influence line gives the magnitude of R_B when the unit load acts at distance x.

* The position of the load influences the value of R_B.

* R_B attains its maximum value when the load is directly over the right support (at B).

* Influence lines can be drawn for reactions, shears and moments in a beam, and member forces in a truss.

* For determinate structures, influence lines are straight lines.

*** Difference ~~between the moment~~ diagram and influence line:**

* A __moment diagram__ gives the value of the bending moment at __different points__ in a beam for a load that is __fixed__ in place.

* An __influence line for moment__ gives the value of the bending moment at only __one point__ in the beam for different locations of the load (__moving load__).

8.3 Influence Lines for Beams:

__Example__ :

$$R_A = \frac{10-x}{10} = 1 - \frac{x}{10} , \qquad 0 < x < 10$$

$$R_B = \frac{x}{10} , \qquad 0 < x < 10$$

When the load is at point C,

at $x = 6$,

$$R_A = 1 - \frac{6}{10} = 0.4 \text{ kN} .$$

$$R_B = 0.6 \text{ KN} .$$

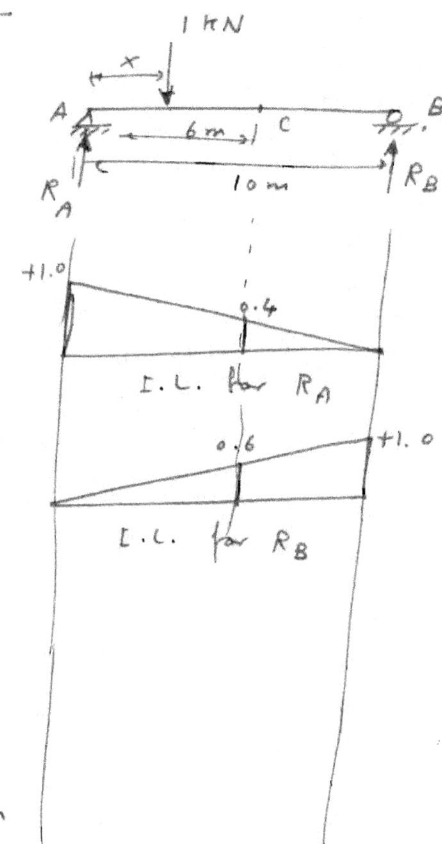

__Notes__ :

① When the load is directly over one of the supports, the reaction at that support resists the entire load.

② As the load moves away from the support, the value of the reaction at the support decreases linearly until the load reaches the other support, at which time the value of the reaction at the first support is zero.

(3) From equilibrium, $\Sigma F_y = 0$:

$$R_A + R_B = 1.0 .$$

The ordinates of the influence lines for the two reactions must add up to unity for any point on the beam.

---

* Draw the I.L. for the bending moment at C.

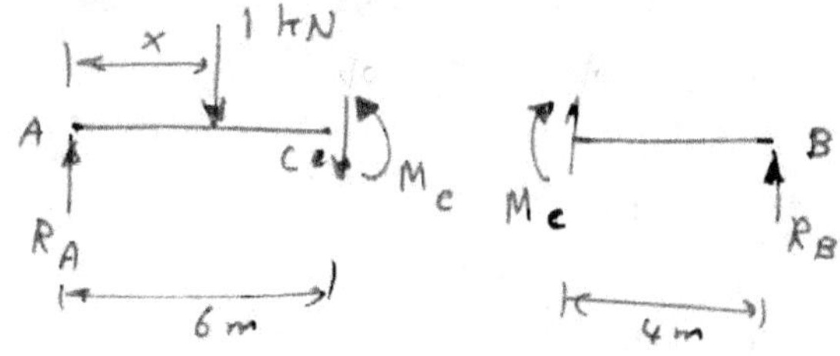

$M_c = 4 R_B$,

$\qquad 0 < x < 6$

but $R_B = \dfrac{x}{0}$

$M_c = \dfrac{2}{5} x$, $0 < x < 6$

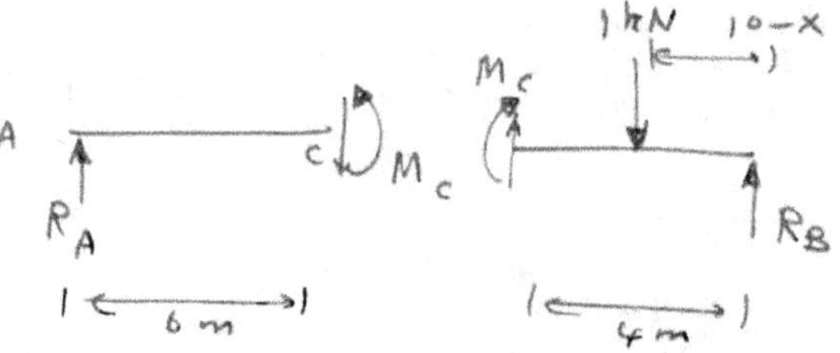

$M_c = 6 R_A$, $\qquad 6 < x < 10$

but $R_A = 1 - \dfrac{x}{10}$

$M_c = 6 - \dfrac{3}{5} x$, $\qquad 6 < x < 10$

at $x = 6$, $M_c = 2.4$ kN·m.

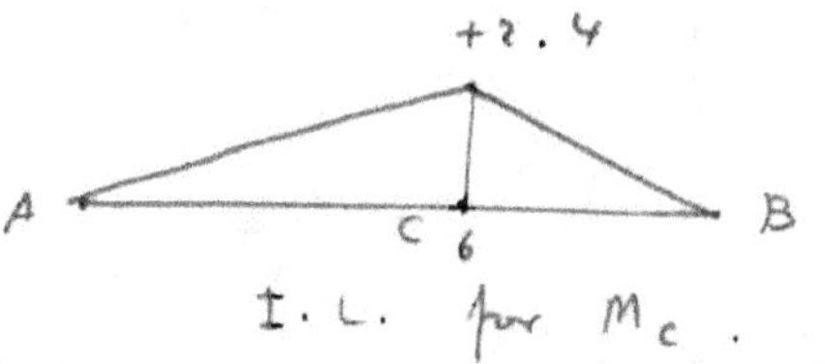

I.L. for M_c.

Draw the I.L for the shear at C :

$V_c = - R_B$, $\qquad 0 < x < 6$

$\qquad = -x/10 .$

$V_c = + R_A$, $\qquad 6 < x < 10$

$\qquad = 1 - x/10 .$

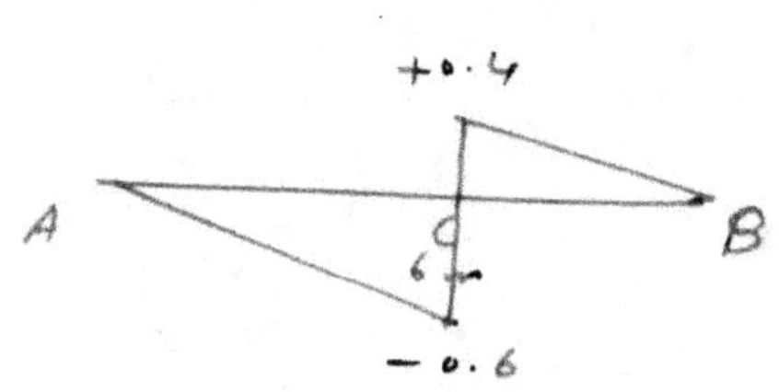

I.L. for V_c.

Example 8.1 :

For the beam shown in the figure, construct the influence lines for the following :

(a) the reactions at A and B.
(b) the moments at A and D.
(c) the shears at D and just to the left of A.

Solution :

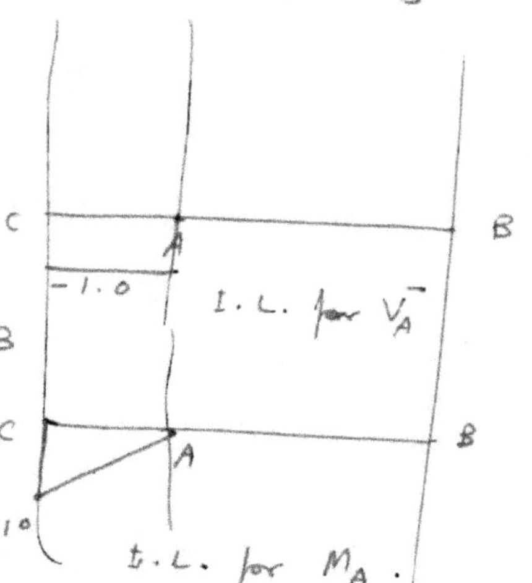

Reactions :

$+\!\!\curvearrowleft \quad \Sigma M_A = 0 :$

$R_B(20) + 1(40) = 0 , \quad R_B = -0.5 \; k .$

$R_A = 1 - (-0.5) = 1.5 \; k \uparrow .$

$R_A + R_B = 1.0$

Shear and Moment at A :

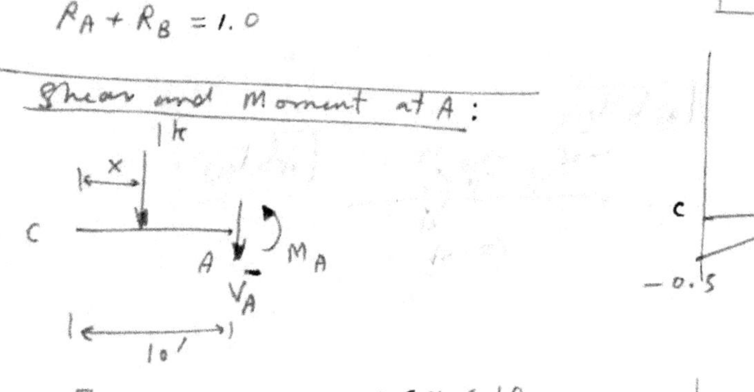

$\overline{V}_A = -1 \; k , \qquad 0 < x < 10$

$M_A = -1(10 - x)$
$\quad = x - 10 , \qquad 0 < x < 10 .$

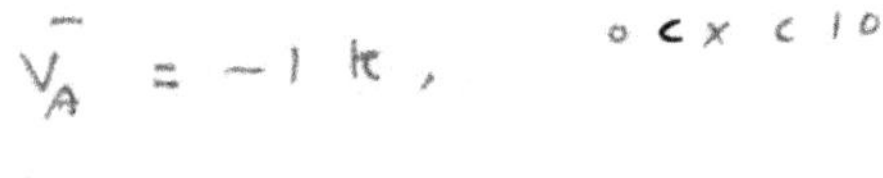

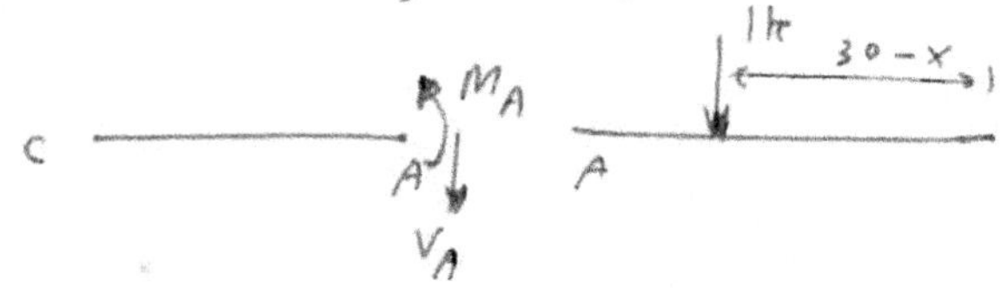

$V_A = 0 , \qquad 10 < x < 30$

$M_A = 0 , \qquad 10 < x < 30$

Shear and Moment at D:

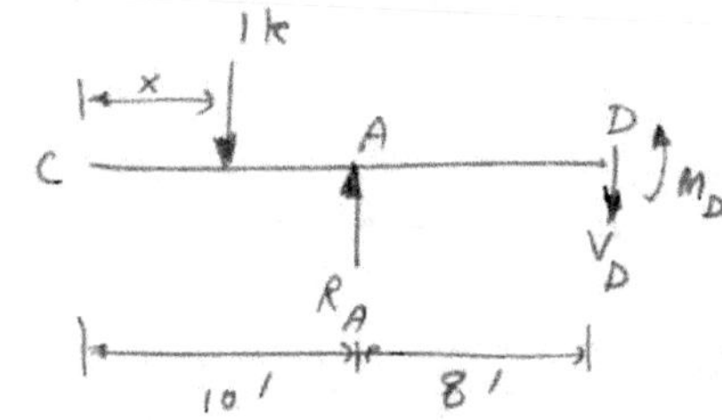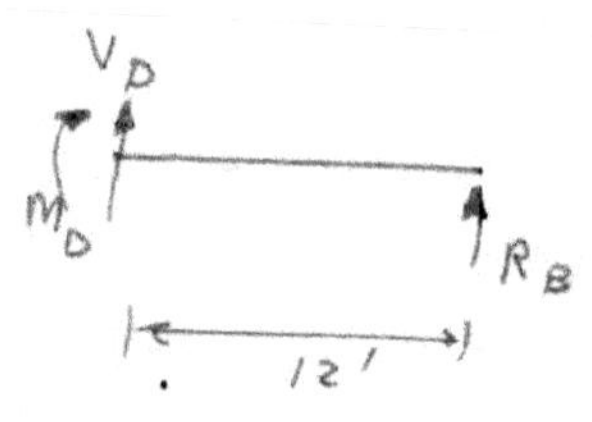

$$V_D = -R_B \quad , \quad 0 < x < 18$$

$$M_D = +12\,R_B \quad , \quad 0 < x < 18$$

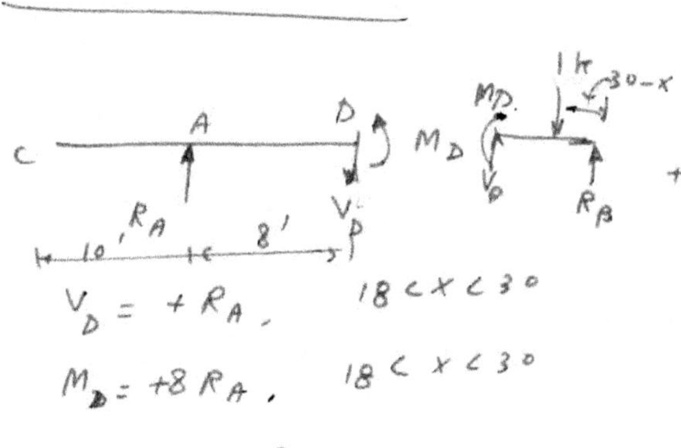

$$V_D = +R_A \quad , \quad 18 < x < 30$$

$$M_D = +8\,R_A \quad , \quad 18 < x < 30$$

at $x = 8$, $\quad \dfrac{R_B}{8} = \dfrac{1}{20}$

$$R_B = \dfrac{8}{20} = 0.4$$

at $x = 8$, $\quad \dfrac{R_A}{12} = \dfrac{1}{20}$

$$R_A = \dfrac{12}{20} = +0.6$$

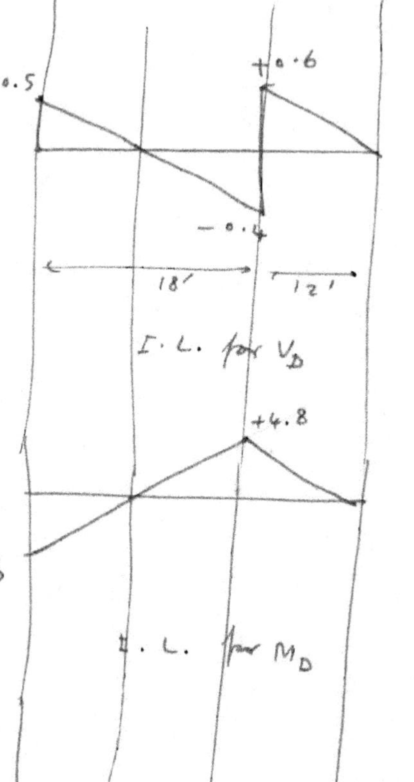

Example 8.2 :

For the beam shown in the figure, construct influence lines

for : (1) the reactions at A, B, and D.

   (2) the moment at B.

   (3) the shear at C.

Solution :

Reactions and Shear at C :

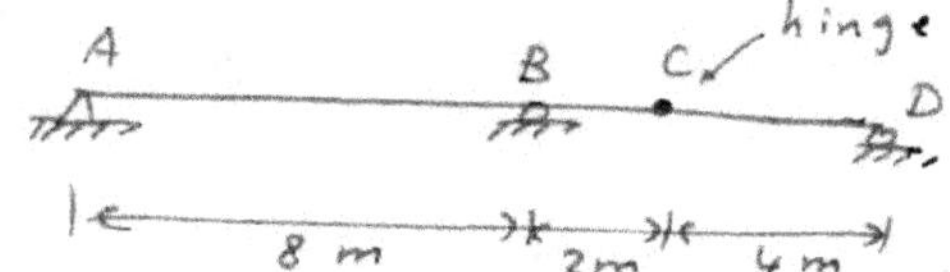

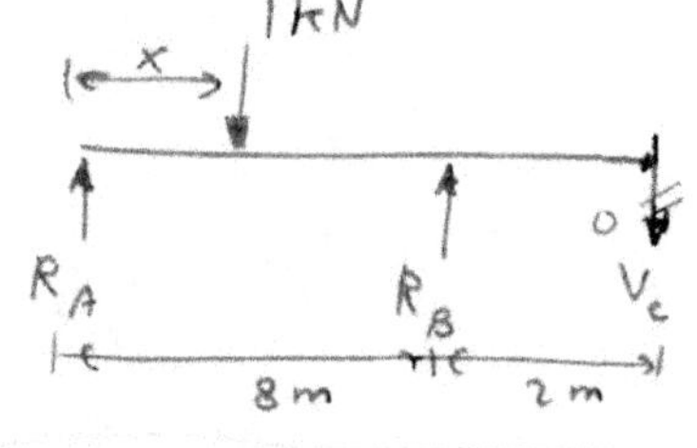

For $0 < x < 10$ m ,

$$R_D = 0 \quad (\Sigma M_c = 0).$$

$$\therefore V_c = 0.$$

$+\circlearrowleft \Sigma M_A = 0 : \quad R_B(8) - 1(x) = 0$

$$R_B = \frac{x}{8}.$$

$$R_A + R_B = 1$$

$$\therefore R_A = 1 - \frac{x}{8}.$$

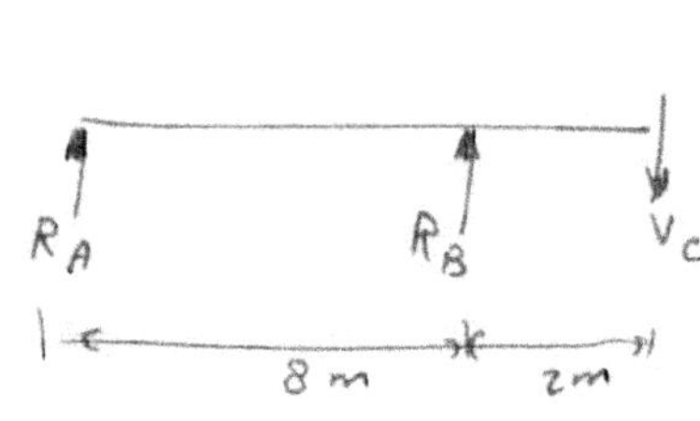

$+\circlearrowleft \Sigma M_c = 0 : \quad R_D(4) - 1[4 - 14 + x] = 0$

$$4R_D + 10 - x = 0$$

$$R_D = \frac{x - 10}{4} , \quad 10 < x < 14$$

$$V_c = 1 - R_D = 1 - \left(\frac{x-10}{4}\right) = \frac{4 - x + 10}{4} = \frac{14 - x}{4}.$$

$\Sigma M_A = 0:$

$$R_B(8) - V_c(10) = 0$$

$$R_B = \frac{10}{8} V_c = \frac{\overset{5}{\cancel{10}}}{\underset{4}{\cancel{8}}}\left(\frac{14-x}{4}\right) = \frac{5}{16}(14-x)$$

$$R_A = V_c - R_B = \frac{14-x}{4} - \frac{5}{16}(14-x)$$

$$= \frac{14-x}{4}\left(1 - \frac{5}{4}\right) = -\frac{1}{4}\left(\frac{14-x}{4}\right) = \frac{x-14}{16}.$$

<u>Moment at B:</u>

① For $\quad 0 < x < 8$

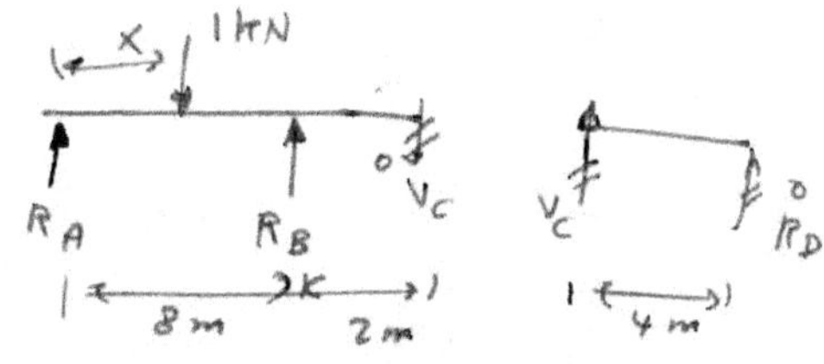

It is clear that $M_B = 0$.

② For $\quad 8 < x < 10$

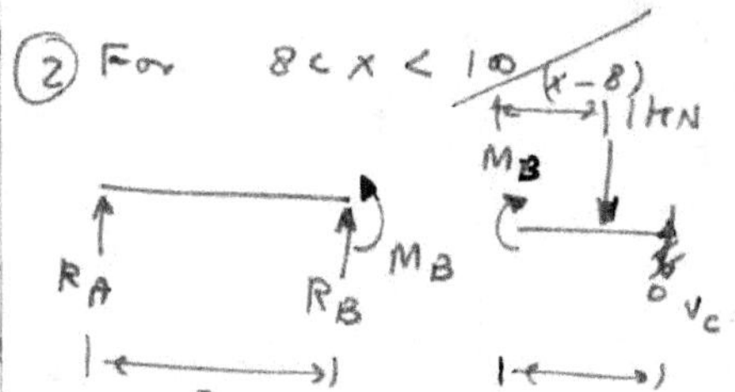

$$M_B = -(x-8) = 8-x.$$

③ For $10 < x < 14$,

$$V_c = \frac{14-x}{4}$$

$$M_B = -2V_c = -\frac{2}{4}(14-x)$$

$$= \frac{x-14}{2}.$$

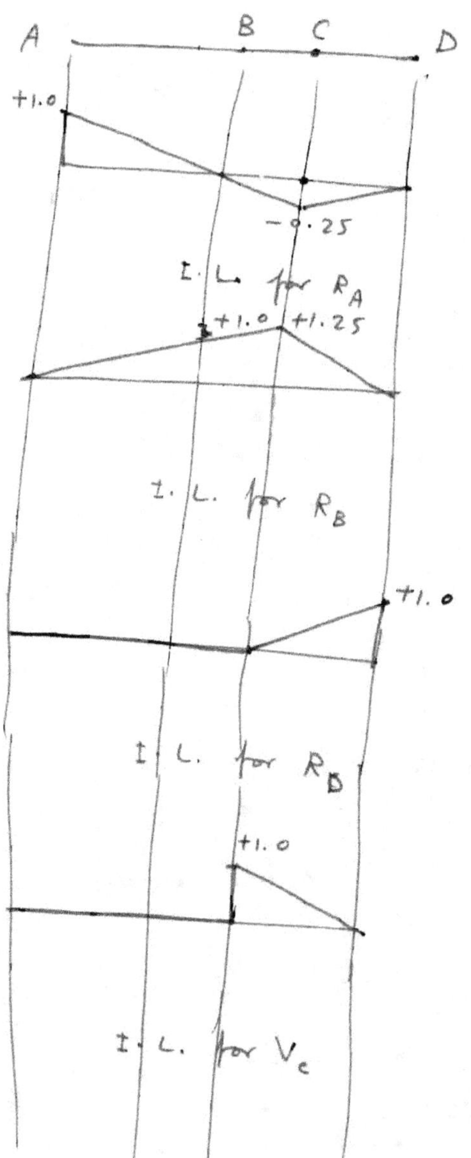

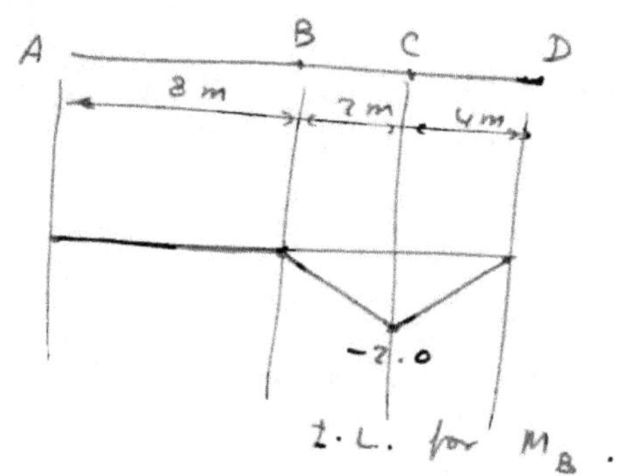

8.4 Influence Lines for Trusses:

* A unit load is moved along the loaded <u>chord</u> of the truss. The load is applied to the truss <u>only</u> at the joints.

* See the <u>bridge truss</u> in the figure.

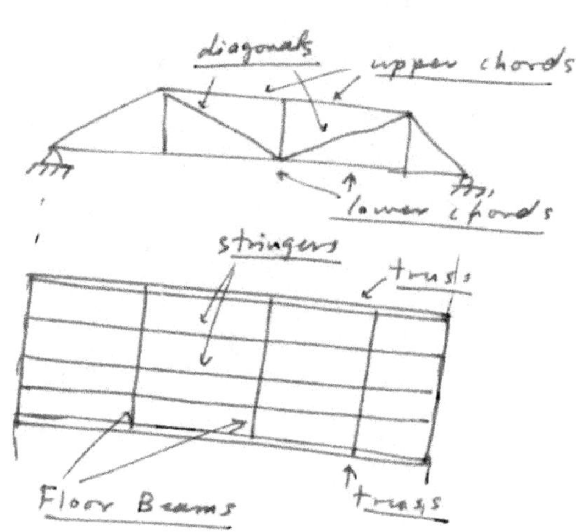

* The steel or concrete <u>deck</u> of a truss bridge rests on <u>stringers</u> (longitudinal members). The stringers span between <u>floor beams</u> (transverse members). The floor beams are supports by the joints of the trusses.

* The load of the vehicles on the bridge is transferred as follows:

  <u>deck → stringers → floor beams → joints of trusses.</u>

* Assume the load to be applied to the truss at the <u>lower chord</u>.

Example :

Assume the moving load is at the lower chord of the truss.

* The influence lines for the reactions at A and G are the same for simply-supported beams.

* Draw the I.L. for member BC.

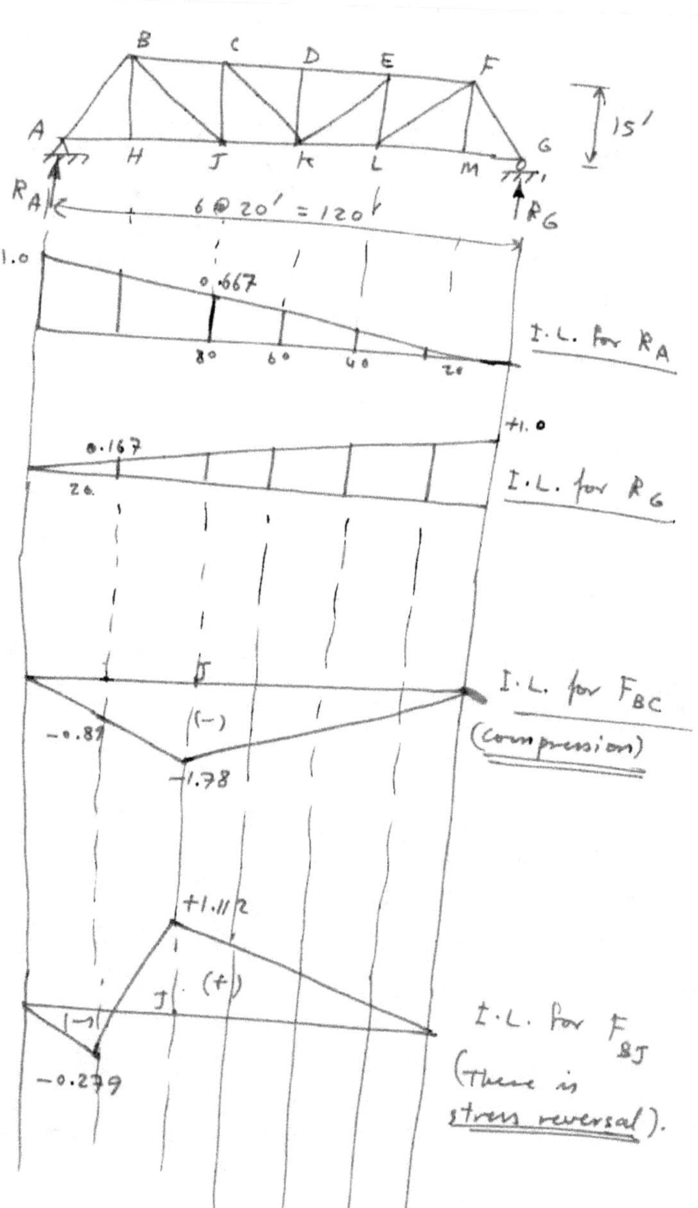

From the R.H.S.,

$+\!\!\!\sum M_J = 0:$

$$F_{BC}(15) + R_G(80) = 0$$

$$\boxed{F_{BC} = -5.33\, R_G}$$

① If the unit load is to the left of J.

② If the unit load is to at J or to the right of J.

$+\!\!\!\sum M_J = 0:$

$$-R_A(40) - F_{BC}(15) = 0$$

$$\boxed{F_{BC} = -2.67\, R_A}$$

* Draw I.L. for F_{BJ}:

From the free-body diagram Ⓐ :

$$\sum F_y = 0: \quad F_{BJ}\left(\frac{15}{25}\right) + R_G = 0$$

$$\Rightarrow \boxed{F_{BJ} = -1.67\, R_G}\ \text{for the unit load to the left of J.}$$

From the free-body diagram $\boxed{B}$:

$$\Sigma F_y = 0: \quad -F_{BJ}\left(\frac{15}{25}\right) + R_A = 0$$

$$\boxed{F_{BJ} = +1.67\, R_A} \quad \text{for the unit load at or to the right of } J.$$

Example 8.3:

Construct influence lines for the forces in members AB, BF, and BG for the truss shown in the figure. The loads are applied to the truss at the joints of the lower chord.

Solution:

First, draw influence lines for the reactions at A and E.

① F_{AB}:

Consider Joint A:

1. the unit load is at A.

$$\boxed{F_{AB} = 0}$$

2. The unit load is at F or to the right of F:

$$\Sigma F_y = 0:$$
$$F_{AB}\left(\frac{10}{18.0}\right) + R_A = 0$$

$$\boxed{F_{AB} = -1.803\, R_A}$$

② F_{BF}:

If the load is at A,
$$F_{BF} = 0.$$
If the load is at G or to the right of G,
$$F_{BF} = 0.$$

If the load is at F,
$$\boxed{F_{BF} = 1.0}$$

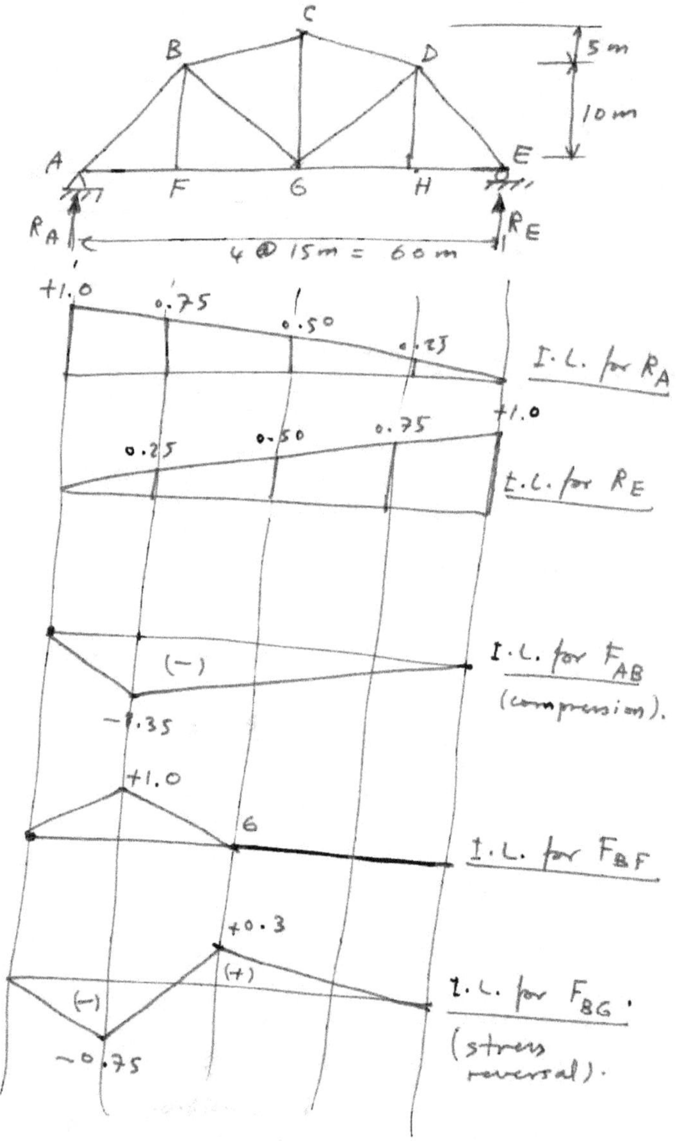

(3) F_{BG} :

$$\frac{10}{x} = \frac{15}{x+15}$$

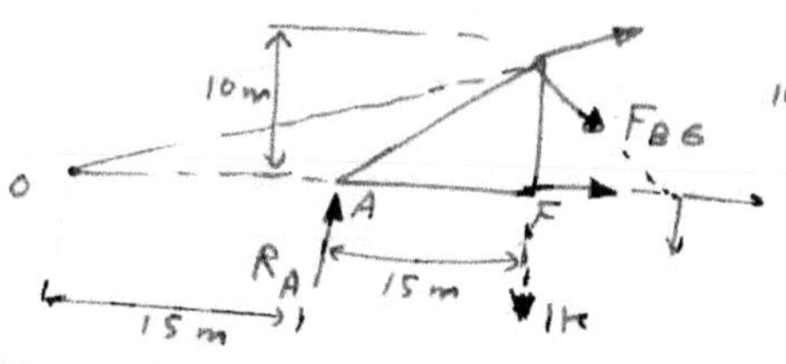

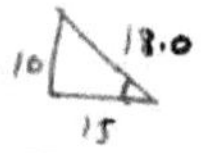

$$15x = 10x + 150$$

$$5x = 150$$

$$x = 30\,m \text{ from joint } F.$$

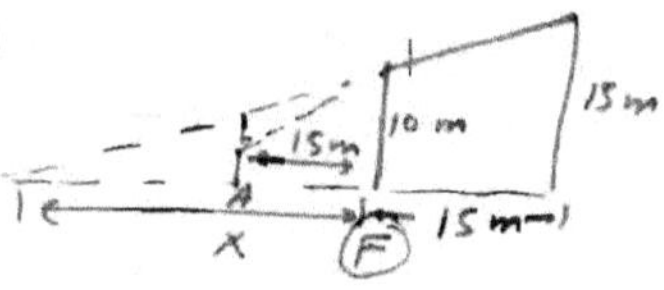

If the unit load is at A,

$+\circlearrowleft \Sigma M_o = 0 :$ $\boxed{F_{BG} = 0}$:

If the unit load is at F,

$+\circlearrowleft \Sigma M_o = 0 :$

$$R_A(15) - 1(30) - F_{BG}\left(\frac{10}{18.0}\right)(45) = 0$$

$$F_{BG} = \frac{18}{450}\left(15 R_A - 30\right)$$

$$\boxed{F_{BG} = 0.6 R_A - 1.2}$$

when $R_A = 0.75$, $F_{BG} = -0.75$

If the unit load is at G or to the right of G :

$+\circlearrowleft \Sigma M_o = 0 :$

$$R_A(15) - F_{BG}\left(\frac{10}{18.0}\right)(45) = 0$$

$$\boxed{F_{BG} = 0.6 R_A}$$

$\boxed{\text{Homework 7}}$ $\underline{8.6}$, $\underline{8.10}$, $\underline{8.14}$, $\underline{8.16}$, $\underline{8.18}$.

Shortcut Method of Drawing Influence Lines for   Beams:

Example:

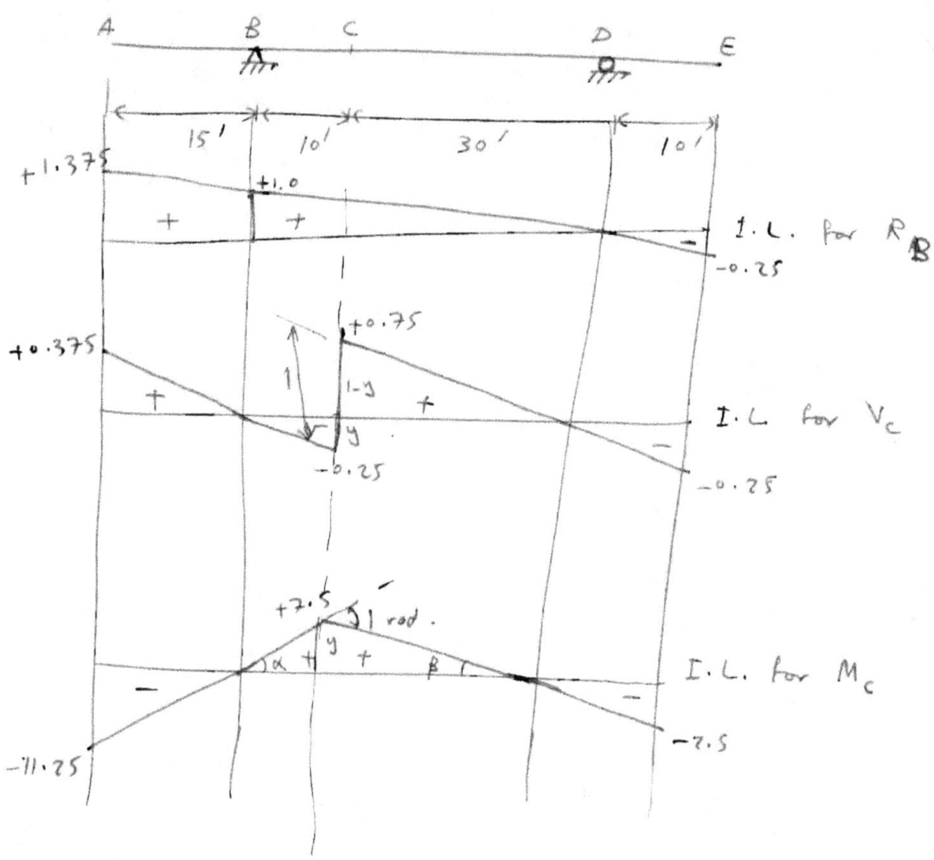

$\underline{V_C}:$ $\dfrac{y}{10} = \dfrac{1-y}{30} \implies 30y = 10 - 10y \implies 40y = 10$

$$y = 0.25$$

$\underline{M_C}:$ $1 \text{ rad.} = \alpha + \beta$

but $\tan\alpha = \dfrac{y}{10} \approx \alpha$ $\Big\}$ for small angles

          $\tan\beta = \dfrac{y}{30} \approx \beta$

$\therefore \dfrac{y}{10} + \dfrac{y}{30} = 1 \implies \dfrac{4y}{30} = 1 \implies y = \dfrac{30}{4} = \underline{7.5}.$

8.5 Uses of Influence Lines:

① Concentrated Loads:

Example:

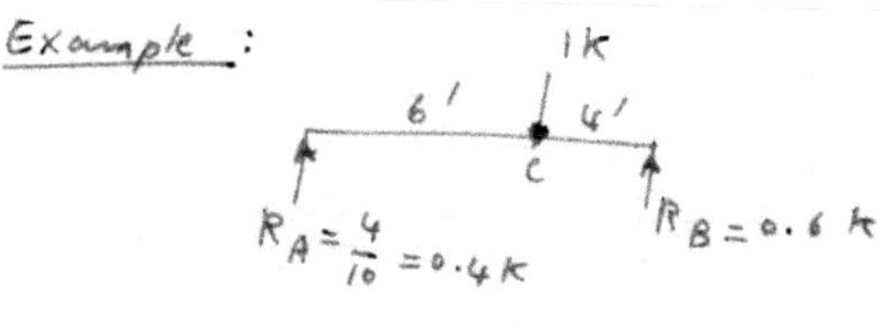

$R_A = \frac{4}{10} = 0.4\,k$ $R_B = 0.6\,k$

$M_c = 6(0.4) = 2.4\ k\text{-ft}.$

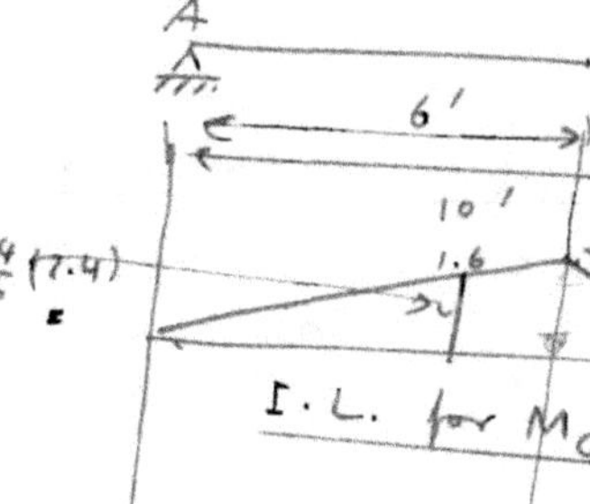

Case (I):

$M_c = 15(2.4) = 36\ k\text{-ft}.$

Case (II):

$M_c = 10(1.6) + 6(2.4) + 8(1.2)$

$\qquad = 40\ k\text{-ft}.$

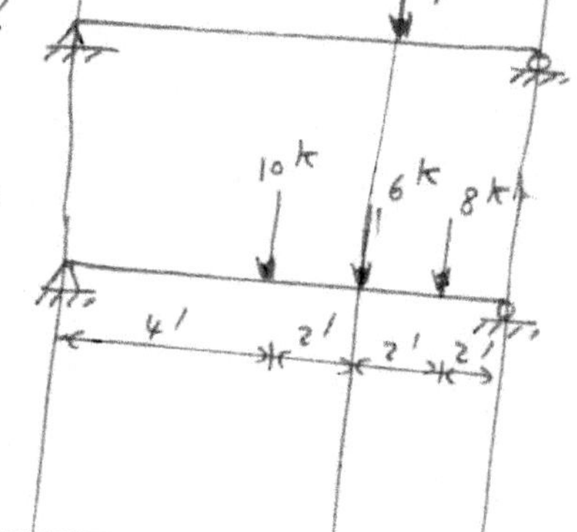

② Uniformly Distributed Loads:

* The value of a **function** (reaction, shear, moment, member force)
due to a uniformly distributed load is equal to the intensity
w of the load multiplied by the area of the influence line
corresponding to the length of beam over which the load is
acting.

Example: U.D.L. , $w = 4\ kN/m$.

$R_A = w \times \text{area}$

$\qquad = 4 \times \frac{1}{2}(1.0)(12)$

$\qquad = 24\ kN.$

Another way, find the resultant concentrated load:

$F_R = 4(12) = 48.$

$R_A = 48(0.5) = 24\ kN.$

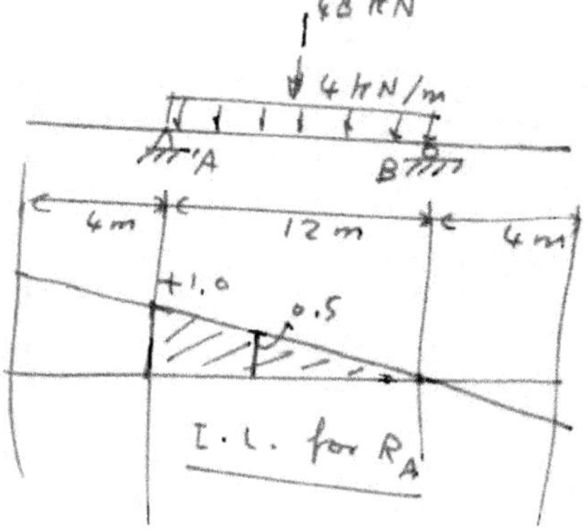

Example :

$$R_A = 4\left[\tfrac{1}{2}(1.333)(16) - \tfrac{1}{2}(0.333)(4)\right]$$

$$= 40\ kN.$$

or resultant load $80\ kN$.

$$R_A = 80\,(0.5) = 40\ kN.$$

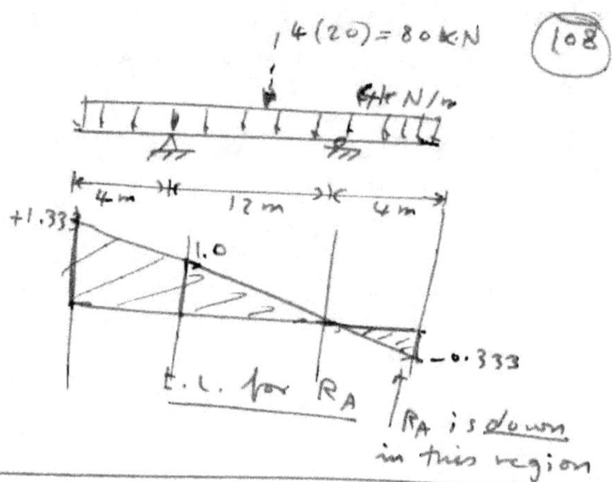

Example 8.4 :

For the beam shown in the figure, find the maximum value of M_D due to a uniform load of intensity $5\ kN/m$, which can act over any part of the beam, and two concentrated loads of $10\ kN$ each, with a fixed distance of $4\ m$ between them.

Solution :

(1) the unit load is to the left of D :

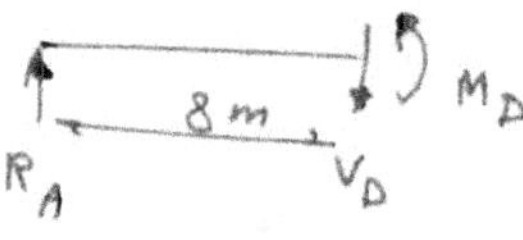

$$\boxed{M_D = 4\,R_B}$$

(2) the unit load is to the right of D :

$$\boxed{M_D = 8\,R_A}$$

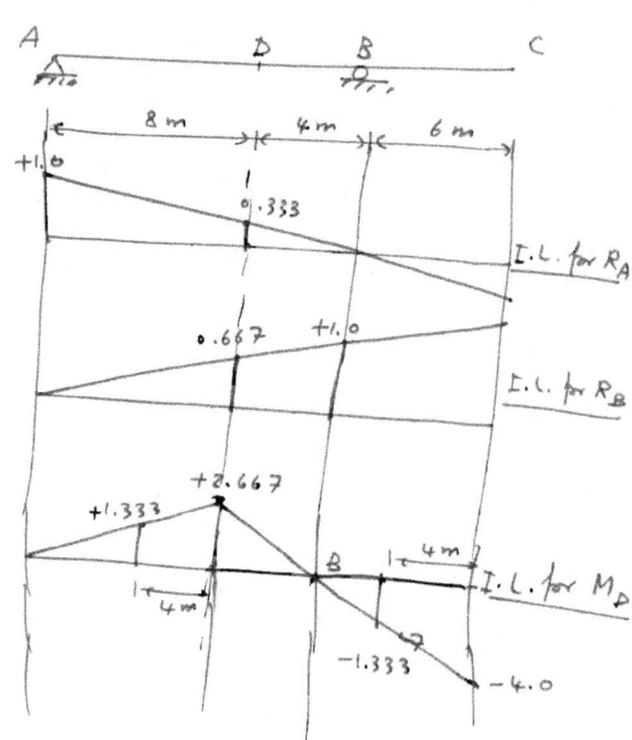

the maximum moment M_D can be <u>positive</u> or <u>negative</u>.

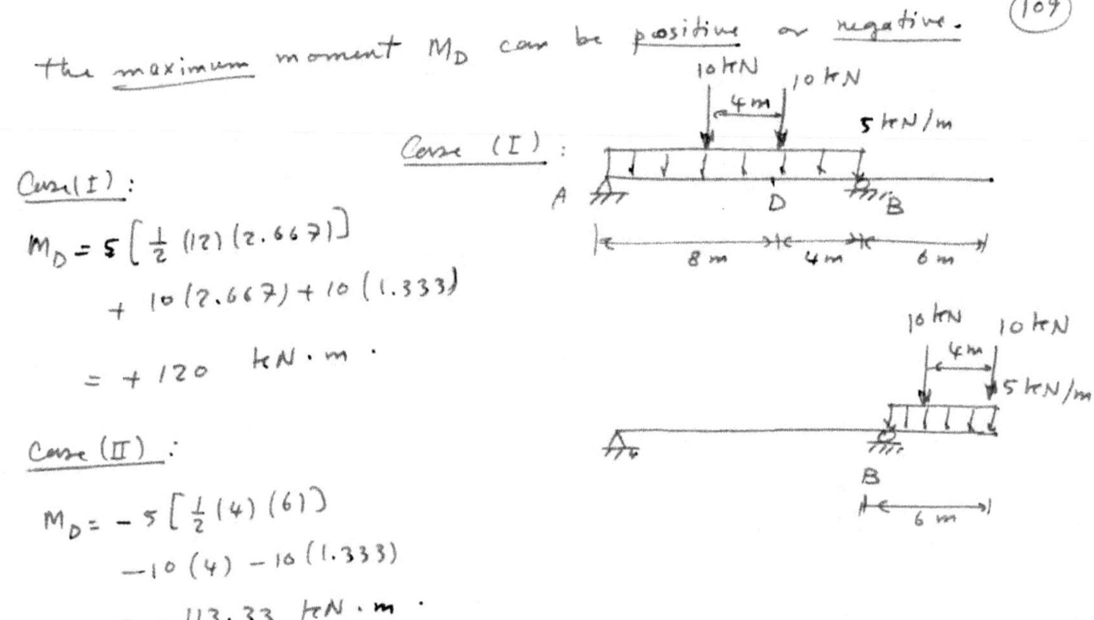

Case (I) :

$$M_D = 5\left[\frac{1}{2}(12)(2.667)\right]$$
$$+ 10(2.667) + 10(1.333)$$

$$= +120 \quad kN \cdot m .$$

Case (II) :

$$M_D = -5\left[\frac{1}{2}(4)(6)\right]$$
$$-10(4) - 10(1.333)$$

$$= -113.33 \quad kN \cdot m .$$

Homework 7   <u>8.6</u> , <u>8.10</u> , <u>8.14</u> , <u>8.16</u> , <u>8.18</u> , <u>8.22</u> .

Influence Lines for Cantilever Beams:

$+\uparrow \Sigma F_y = 0:$

$\qquad R_A - 1 = 0$

$\qquad \Rightarrow \quad R_A = 1.$

$+\circlearrowleft \Sigma M_A = 0:$

$\qquad -M_A - 1(10-x) = 0$

$\qquad M_A = -(10-x), \qquad 0 \le x \le 10.$

Case (1): Load to the right of C:

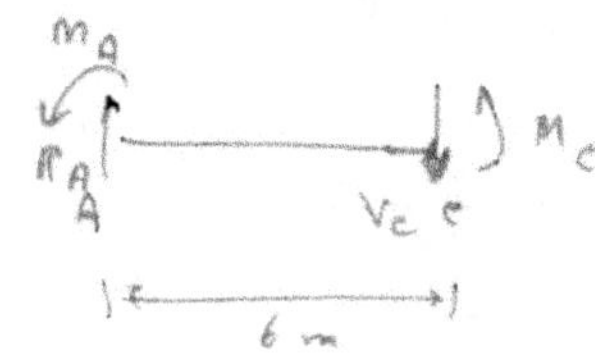

$+\uparrow \Sigma F_y = 0: \quad -V_c + R_A = 0$

$\qquad V_c = R_A$

$+\circlearrowleft \Sigma M_c = 0:$

$\qquad M_c + M_A - R_A(6) = 0$

$\qquad M_c = 6R_A - M_A.$

Case (2): Load to the left of C:

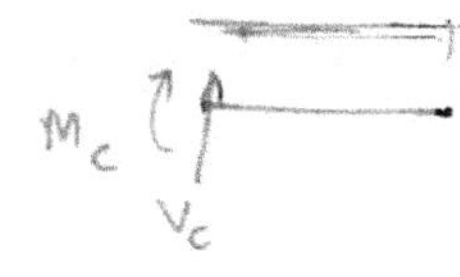

$\qquad V_c = 0$

$\qquad M_c = 0.$

Influence Lines for Girders with Floor Beams:

Example:

Consider the girder with four panels as shown in the figure.

Solution is left as an exercise.

Note that the shear is constant in each panel.

* Loads are applied to the girder only at the panel points.

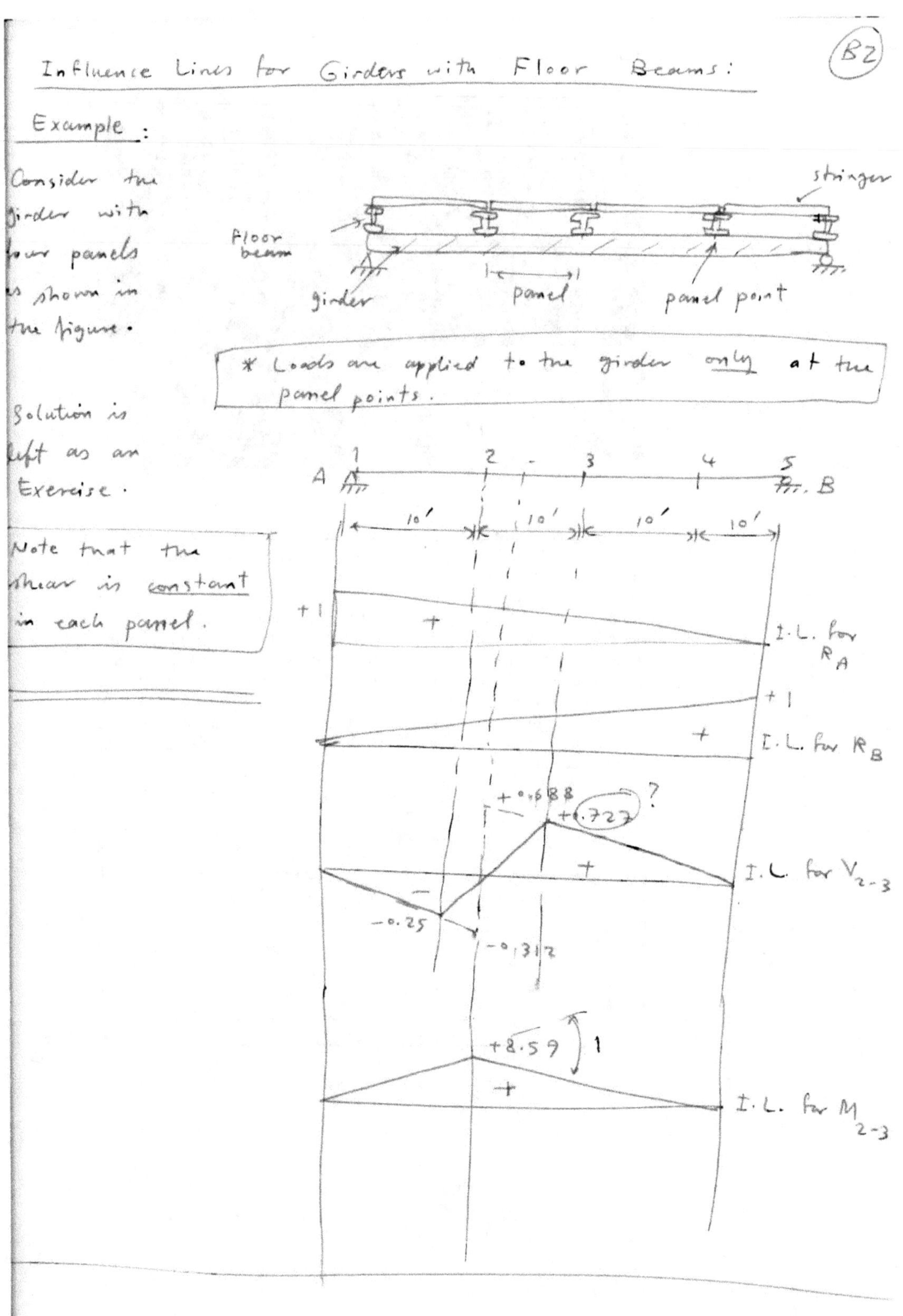

Figure 6.1   Movable crane girder showing motor-driven lifting hoist. Photo taken by author.

6.4*b* is a general FBD from which it is evident, by use of statics, that $R_L = 1 \text{ kip}(L - x)/L$ and $R_R = 1 \text{ kip}(x/L)$; that is, each reaction varies linearly due to the unit moving load. Therefore, all reaction values shown on Fig. 6.4*a* are joined by a straight line. Since Fig. 6.4*a* shows how the position of the unit load "influences" the value of the left reaction, Fig. 6.4*a* is called the influence line for the left reaction of the beam. In a similar manner, the influence line for the right reaction (R_R) can easily be determined with ordinate values as shown in Fig. 6.4*c*. The positive sense shown on the influence line of Fig. 6.4*c* indicates that the direction of R_R is upward. Since these reactions are the shears at the beam support ends, the influence lines of Figs. 6.4*a* and 6.4*c* are influence lines for shear. One can now continue to study influence lines for shear at interior beam locations.

Figure 6.2

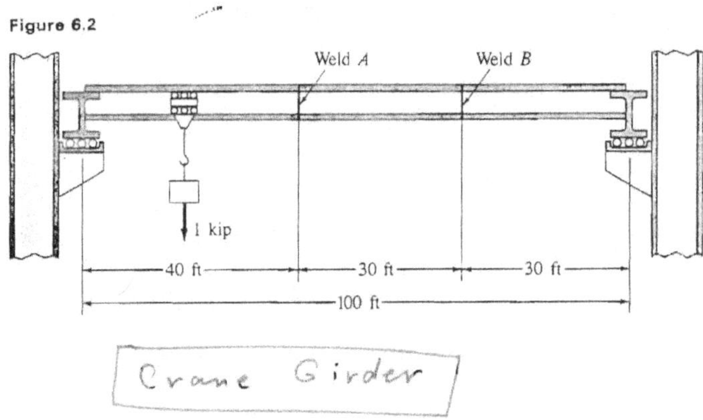

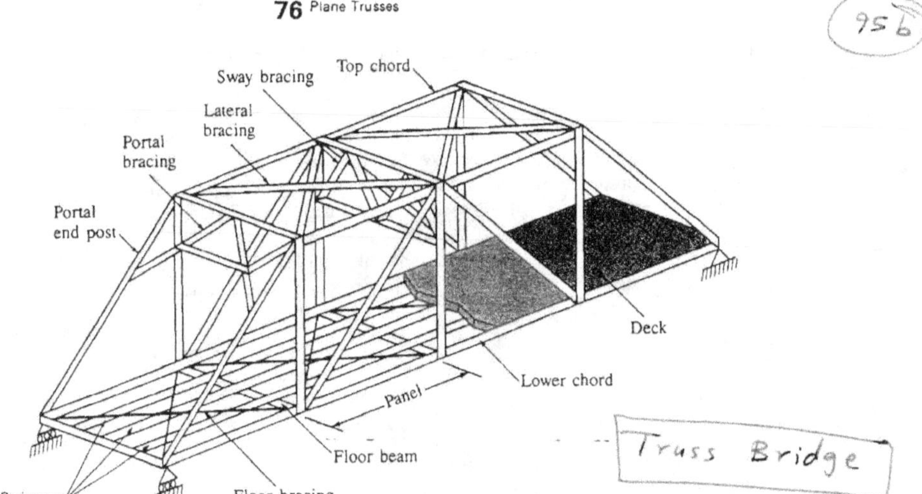

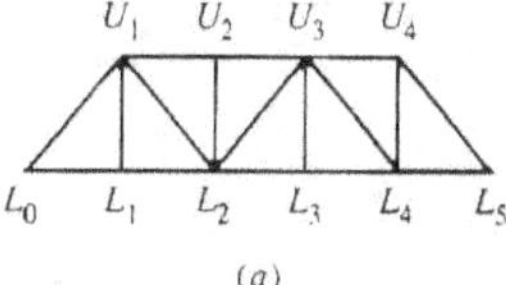

Figure 3.14

3.4. SIGN CONVENTION

Assume *all* truss members act in tension (T). Therefore, when a member force is calculated to be of positive sign, it means that the original tension direction is correct; if negative, the member acts in compression (C).

3.5. TRUSS NOTATIONS

Two commonly used methods of notation are mentioned. First, in modeling trusses for computer solutions, it is customary to define truss joints (nodes) by integer values starting with 1. For classical hand calculation methods, upper chord joints are noted by U_n and lower chord joints by L_n, where n represents a joint number starting with 0. Figure 3.15a demonstrates the later notation where it is observed that upper chord members are defined as $U_n U_{n+1}$ and lower chord members as $L_n L_{n+1}$; vertical members are noted as $U_n L_n$ (e.g., $U_3 L_3$), and diagonal members as either $U_n L_{n+1}$ or $L_n U_{n+1}$ (e.g., $U_1 L_2$ or $L_2 U_3$). Figure 3.15b contains no vertical members and shows that joint numbers at both upper and lower chords progress by $n + 2$. It is also common to specify truss joints by letter notation (see Fig. 3.17).

Figure 3.15

(a) (b)

6.6. BRIDGE DECK AND BUILDING FLOOR SYSTEMS

In addition to studying isolated beams, it must be realized that live loads typically are applied to systems of beams and girders that comprise either bridge decks or building floor systems. In order to facilitate the task of developing the influence lines for these floor members, it is appropriate to first study how the moving loads are transferred and supported in each system.

Figures 6.17a and b show a simple arrangement typically found in bridge decks that is comprised of a deck slab, stringers, floor beams, and girders.

Figure 6.17

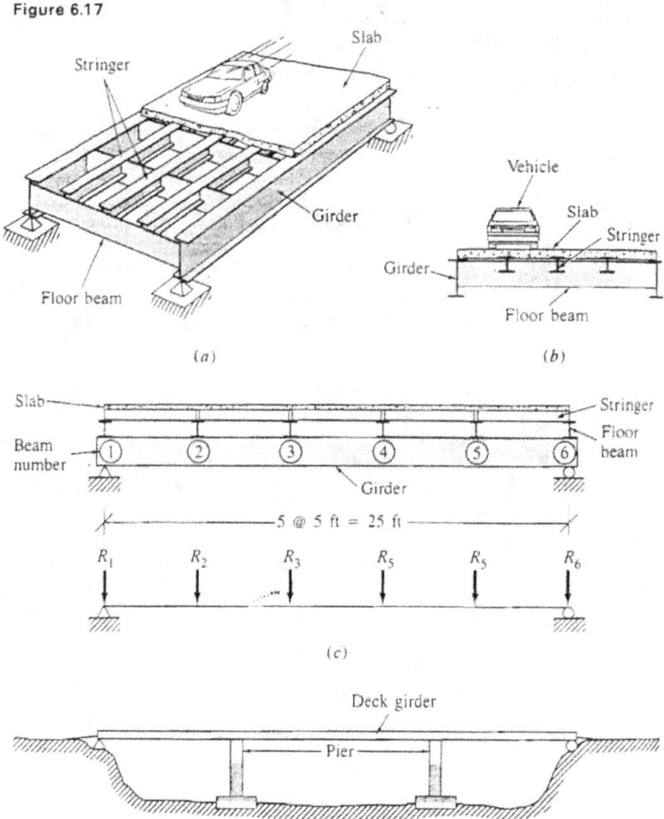

Chapter 9
Arches and Cables (not required)

9.1 Arches:

* The arch is a structural system in which the primary internal force is _axial compression_.

* Arches are able to span much larger distances than beams due to their reduced bending moments.

Example:

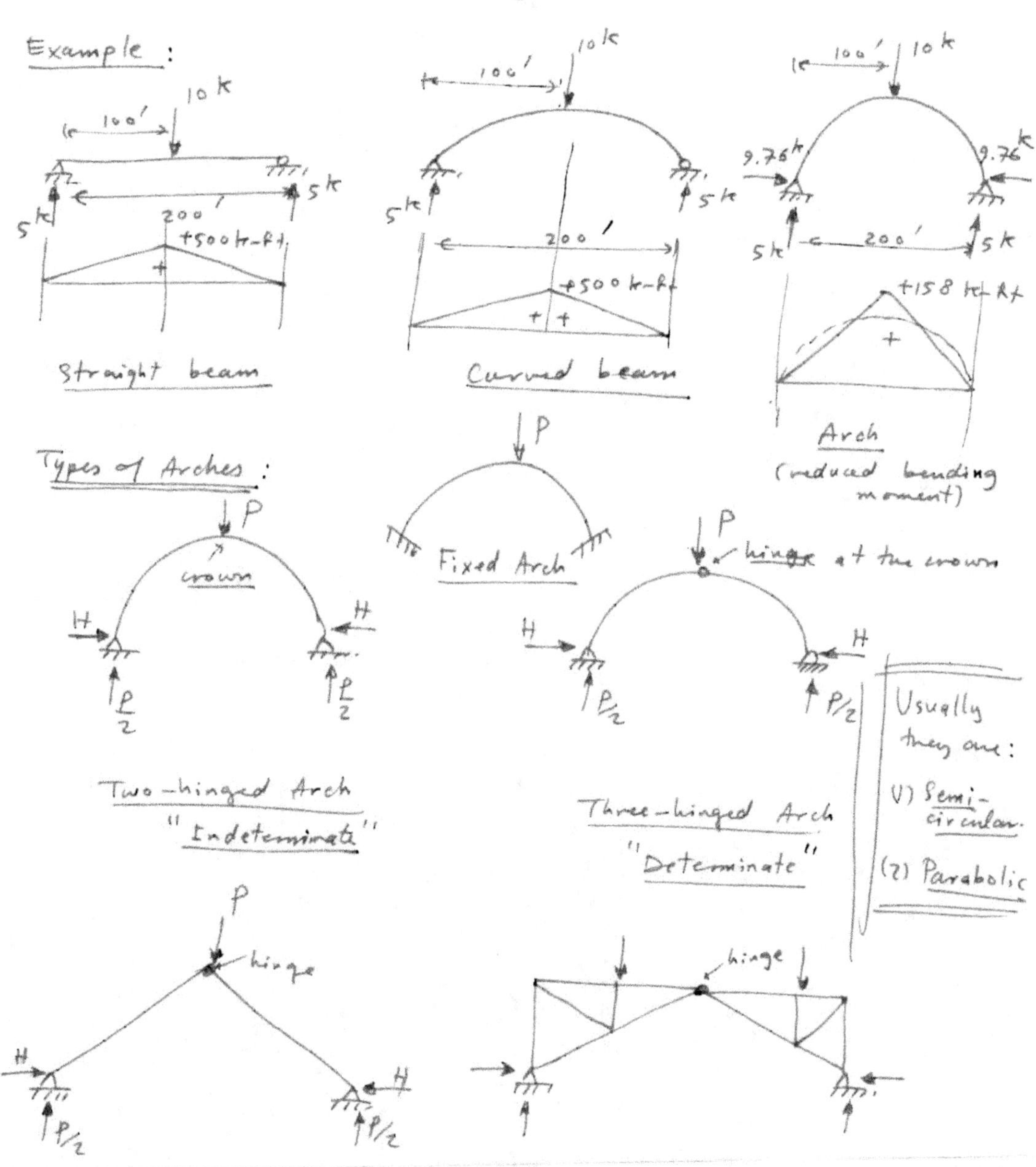

Types of Arches:

Example 9.1 :

Determine the reactions for the arch shown in the figure.

Solution :

$+\circlearrowleft \Sigma M_A = 0$:

$V_B(40) - 60(28) - 100(10) = 0$

$V_B = 67\ kN \uparrow$

$+\uparrow \Sigma F_y = 0$:

$V_A + 67 - 100 - 60 = 0$

$V_A = 93\ kN \uparrow$.

Consider the left-hand part :

$+\circlearrowleft \Sigma M_C = 0$:

$100(10) + H_A(10) - 93(20) = 0$

$H_A = 86\ kN \longrightarrow$

For the entire arch :

$\rightarrow \Sigma F_x = 0$: $H_B = 86\ kN \longleftarrow$.

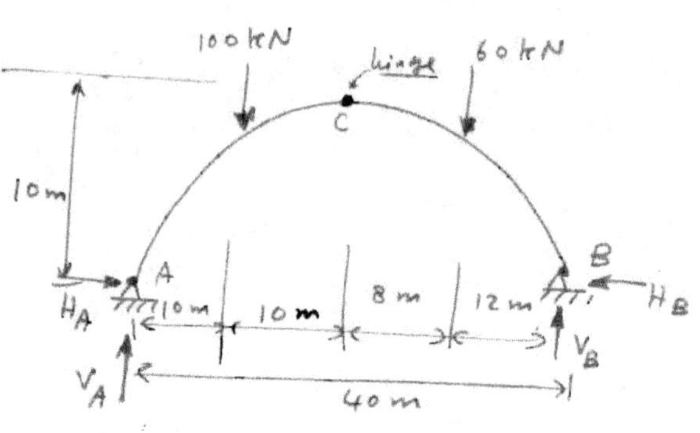

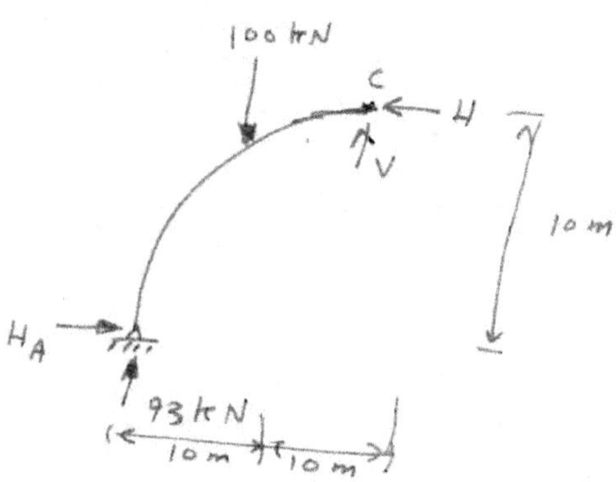

Example 9.2 :

Determine the reactions and construct the moment diagram for the arch shown in the figure.

Solution :

$+\circlearrowleft \Sigma M_A = 0$:

$+ V_B(100) + H_B(5)$

$- 30(72) - 40(20) = 0$

$\therefore 100 V_B + 5 H_B = 2960$

$\underline{\hspace{2cm}}$ (1)

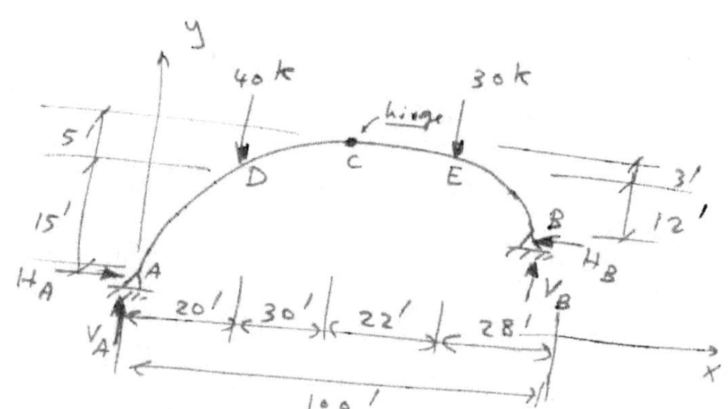

Consider the right-hand part:

$+\circlearrowleft \Sigma M_c = 0$:

$$V_B(50) - H_B(15) - 30(22) = 0$$

$$50 V_B - 15 H_B = 660 \qquad \text{——— (2)}$$

Solve equations (1) & (2) simultaneously:

$$+300 V_B + 15 H_B = +8880$$
$$50 V_B - 15 H_B = 660$$

$$350 V_B = 9540 \qquad \Rightarrow \qquad V_B = 27.3 \text{ k} \uparrow.$$

$$\therefore \ H_B = \frac{1}{15}\left[50(27.3) - 660\right] = 46.9 \text{ k} \leftarrow.$$

For the entire arch:

$\xrightarrow{+} \Sigma F_x = 0$: $\quad H_A = 46.9 \text{ k} \longrightarrow.$

$+\uparrow \Sigma F_y = 0$: $\quad V_A + 27.3 - 40 - 30 = 0 \qquad \Rightarrow V_A = 42.7 \text{ k} \uparrow.$

Bending Moment :

$\curvearrowleft +)\ \Sigma M_D = 0:$

$M_D + 46.9\,(15)$
  $- 42.7\,(20) = 0$

$M_D = 150.5\ \text{k-ft.}$

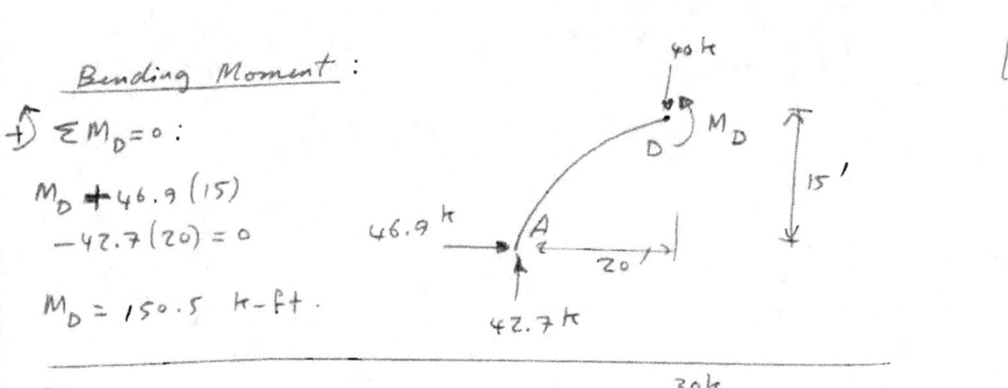

$\curvearrowleft +)\ \Sigma M_E = 0:$

$-M_E - 46.9\,(12) + 27.3\,(28) = 0$

$M_E = 201.6\ \text{k-ft.}$

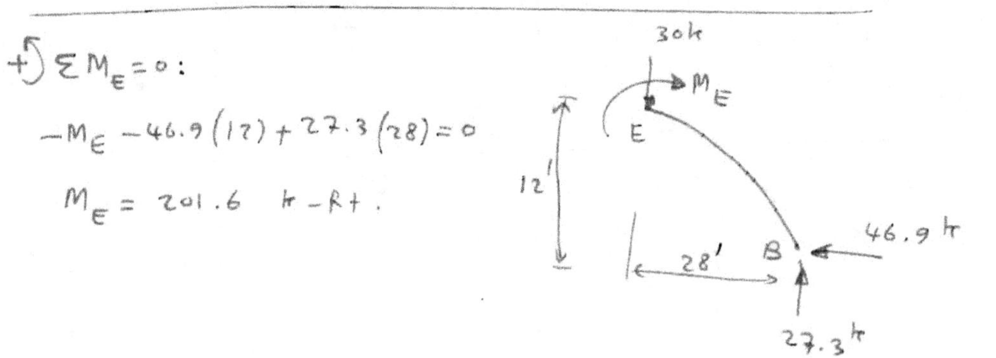

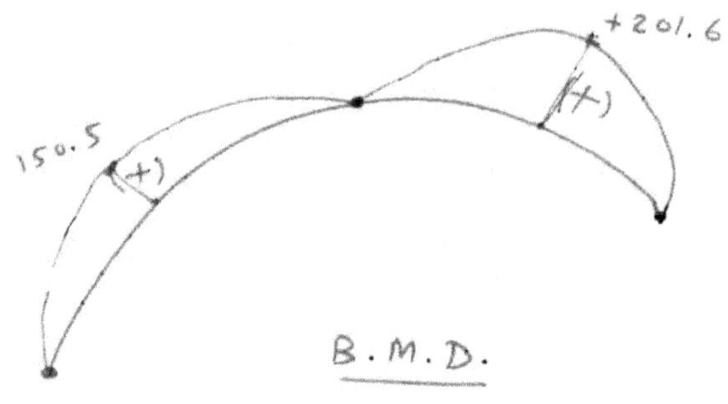

B.M.D.

Example: The three-hinged parabolic arch shown in the figure has a span of 100 m with two supports at the same level. The two supports are hinged and a third hinge is located at the crown, with a rise of 20 m. Find the normal force, shear, and moment diagrams in the arch for a uniform load of 30 kN/m acting over the horizontal projection of the arch.

Solution:

(1) Calculate the reactions:

Due to symmetry,

$$V_A = V_B = \frac{1}{2}(30 \times 100)$$
$$= 1500 \text{ kN} \uparrow.$$

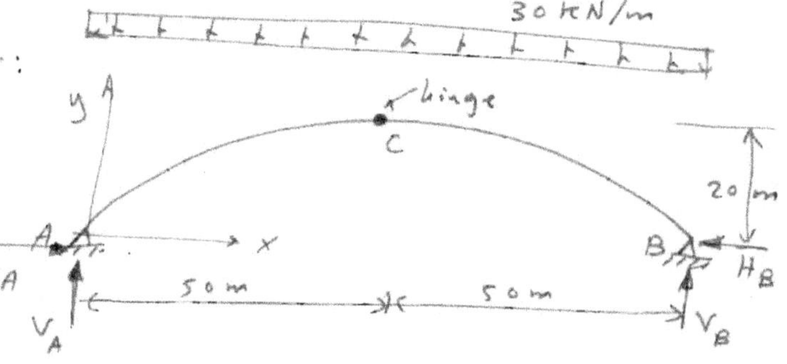

Consider the right-hand part:

$$\circlearrowleft \quad \Sigma M_C = 0:$$

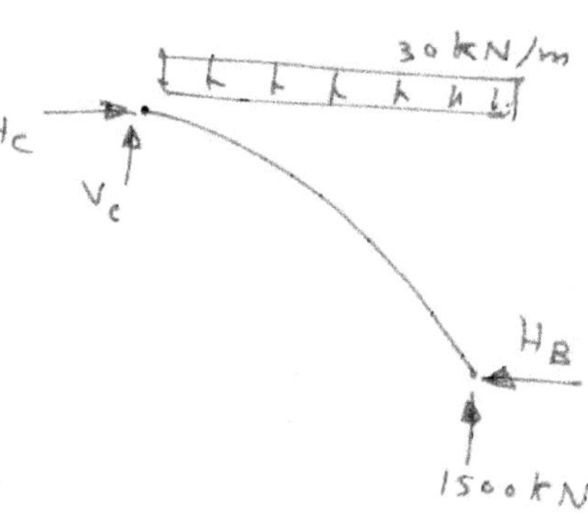

$$-30(50)\left(\frac{50}{2}\right) - H_B(20) + 1500(50) = 0$$

$$H_B = 1875 \text{ kN} \leftarrow$$

$$\therefore \quad H_A = 1875 \text{ kN} \longrightarrow.$$

(2) Calculate the internal forces (normal force, shear, & moment):

First find the equation of the parabola:
$$y = a + bx + cx^2$$

point A: $x = 0$, $y = 0$ $\Longrightarrow$ $a = 0$.

point B: $x = 100$, $y = 0$ $\Longrightarrow$ $0 = 100b + (100)^2 c$ —————— (1)

point C: $x = 50$, $y = 20$ $\Longrightarrow$ $20 = 50b + (50)^2 c$ —————— (2)

Solve eqs. (1) & (2) simultaneously for a, b:

From eq (1): $b = -100c$.

$$\therefore \quad 20 = -50(100)\,c + (50)^2\,c$$

$$20 = 50c\,(50-100)$$

$$20 = -(50)^2\,c \implies c = -\frac{20}{2500} = -0.008$$

$$b = -100\,c = 0.8$$

$$\therefore \quad \boxed{y = 0.8x - 0.008\,x^2}$$

Slope $\quad \boxed{\dfrac{dy}{dx} = \tan\phi = 0.8 - 0.016\,x}$

Consider the section shown in the figure:

$\xrightarrow{+} \Sigma F_x = 0:$

$$N\cos\phi + V\sin\phi + 1875 = 0 \quad\text{——— (1)}$$

$+\uparrow \Sigma F_y = 0:$

$$N\sin\phi - V\cos\phi$$
$$+1500 - 30x = 0 \quad\text{——— (2)}$$

Solve eqs. (1) & (2) simultaneously:

$$N\cos^2\phi + V\sin\phi\cos\phi + 1875\cos\phi = 0 \quad\text{——— (1)}$$

$$N\sin^2\phi - V\sin\phi\cos\phi + 1500\sin\phi - 30x\sin\phi = 0 \quad\text{——— (2)}$$

$$\rule{12cm}{0.4pt}$$

$$N + 1875\cos\phi + 1500\sin\phi - 30x\sin\phi = 0$$

$$\therefore \quad N = 30x\sin\phi - 1500\sin\phi - 1875\cos\phi$$

$$\text{but} \quad \sin\phi = \frac{\tan\phi}{\sqrt{1+\tan^2\phi}} = \frac{0.8 - 0.016x}{\sqrt{1+(0.8-0.016x)^2}}$$

$$\cos\phi = \frac{1}{\sqrt{1+\tan^2\phi}} = \frac{1}{\sqrt{1+(0.8-0.016x)^2}}$$

$$\therefore \quad N = \frac{30x\,(0.8-0.016x)}{\sqrt{1+(0.8-0.016x)^2}} - \frac{1500\,(0.8-0.016x)}{\sqrt{1+(0.8-0.016x)^2}} - \frac{1875}{\sqrt{1+(0.8-0.016x)^2}}$$

$$\therefore \quad N(x) = \frac{30x(0.8 - 0.016x) - 1500(0.8 - 0.016x) - 1875}{\sqrt{1 + (0.8 - 0.016x)^2}}$$

$$= \frac{24x - 0.48x^2 - 1200 + 24x - 1875}{\sqrt{1 + (0.8 - 0.016x)^2}}$$

$$\boxed{N(x) = \frac{-0.48x^2 + 48x - 3075}{\sqrt{1 + (0.8 - 0.016x)^2}}} \quad , \quad 0 \le x \le 100$$

Multiply eq. (1) by $-\sin\phi$ and (2) by $\cos\phi$:

$$-N\sin\phi\cos\phi - V\sin^2\phi - 1875\sin\phi = 0$$
$$N\sin\phi\cos\phi - V\cos^2\phi + 1500\cos\phi - 30x\cos\phi = 0$$
$$\overline{\quad\quad\quad\quad\quad\quad\quad\quad\quad\quad\quad\quad\quad\quad\quad}$$
$$-V - 1875\sin\phi + 1500\cos\phi - 30x\cos\phi = 0$$

$$V = \frac{-1875(0.8 - 0.016x)}{\sqrt{1 + (0.8 - 0.016x)^2}} + \frac{1500}{\sqrt{1 + (0.8 - 0.016x)^2}} - \frac{30x}{\sqrt{1 + (0.8 - 0.016x)^2}}$$

$$= \frac{-1500 + 30x + 1500 - 30x}{\sqrt{1 + (0.8 - 0.016x)^2}} = 0 .$$

$$\therefore \boxed{V(x) = 0} .$$

$+\circlearrowleft \ \Sigma M_x = 0 :$

$$M(x) + 30x\left(\frac{x}{2}\right) + 1875y - 1500x = 0 \quad , \text{ but } y = 0.8x - 0.008x^2$$

$$\therefore M(x) + 15x^2 + 1875(0.8x - 0.008x^2) - 1500x = 0$$

$$M(x) + 15x^2 + 1500x - 15x^2 - 1500x = 0$$

$$\boxed{M(x) = 0}$$

<u>Example</u> : Draw the normal force, shear, and moment diagrams for the <u>three-hinged semi-circular</u> arch shown in the figure under its own weight. The weight of the arch is 2 k/ft of the centerline. The centerline of the arch has a radius of 50 ft.

<u>Solution :</u>
(1) <u>Calculate the reactions :</u>
Due to symmetry,

$$V_A = V_B = \frac{1}{2}(w \cdot \pi r)$$

$$= \frac{1}{2}(2)(\pi)(50)$$

$$= 50\pi$$

$$= 157 \ k.$$

Consider the right-hand-part :

$$dP = w\,ds \quad , \quad \text{but} \quad ds = r\,d\theta$$

$$dP = w\,r\,d\theta$$

$+\circlearrowright \ \Sigma M_c = 0$:

$$157(50) - H_B(50) - \int_0^{\pi/2} dP \cdot (50\cos\theta) = 0$$

$$\cancel{50}(157 - H_B) = \int_0^{\pi/2} w\,r\,d\theta \cdot \cancel{50}(\cos\theta)$$

$$157 - H_B = \int_0^{\pi/2} (2)(50)(\cos\theta)\,d\theta$$

$$157 - H_B = 100 \cdot \Big[\sin\theta\Big]_0^{\pi/2}$$

$$157 - H_B = 100\,(1) \quad \Longrightarrow \quad H_B = 57 \ k \leftarrow$$

$$H_A = \ +57 \ k \longrightarrow .$$

(2) Calculate the internal forces (normal force, shear, & moment):

Consider the section shown in the figure:

$$ds = r\,d\theta$$

$$dP = w\,ds = w\,r\,d\theta$$

$$= 2(50)\,d\theta = 100\,d\theta$$

$\xrightarrow{+}\ \Sigma F_x = 0:$

$$-N_\phi \sin\phi - 57 + V_\phi \cos\phi = 0 \qquad\qquad (1)$$

$+\uparrow\ \Sigma F_y = 0:$

$$N_\phi \cos\phi + V_\phi \sin\phi - \int_0^\phi 100\,d\theta + 157 = 0 \qquad (2)$$

but $\displaystyle \int_0^\phi 100\,d\theta = 100\big[\theta\big]_0^\phi = 100\,\phi$

$\therefore$

$\cos\phi$ | $-N_\phi \sin\phi + V_\phi \cos\phi - 57 = 0 \qquad\qquad (1)$

$\sin\phi$ | $N_\phi \cos\phi + V_\phi \sin\phi - 100\phi + 157 = 0 \qquad (2)$

Solve the two equations simultaneously for N_ϕ, V_ϕ:

$$-N_\phi \sin\phi \cos\phi + V_\phi \cos^2\phi - 57\cos\phi = 0$$

$$N_\phi \sin\phi \cos\phi + V_\phi \sin^2\phi - 100\phi \sin\phi + 157 \sin\phi = 0$$

$$V_\phi - 57\cos\phi - 100\phi \sin\phi + 157 \sin\phi = 0$$

$$V_\phi = 100\phi \sin\phi - 157 \sin\phi + 57\cos\phi$$

$\therefore$ $\boxed{V_\phi = 100\sin\phi\,(\phi - 1.57) + 57\cos\phi}$, $\quad 0 \le \phi \le \pi$

multiply eq.(1) by $-\sin\phi$ and (2) by $\cos\phi$:

$$N_\phi \sin^2\phi - V_\phi \sin\phi \cos\phi + 57 \sin\phi = 0$$

$$N_\phi \cos^2\phi + V_\phi \sin\phi \cos\phi - 100\phi \cos\phi + 157 \cos\phi = 0$$

$$N_\phi + 57\sin\phi - 100\phi \cos\phi + 157 \cos\phi = 0$$

$$\therefore \boxed{N_\phi = 100\cos\phi\,(\phi - 1.57) - 57\sin\phi} \quad , \quad 0 \le \phi \le \pi \qquad \fbox{117}$$

f) $\Sigma M_x = 0$:

$$-\int_0^\phi dP.(50)(\cos\theta - \cos\phi) - 57\left(50\frac{\sin\phi}{\phi} + 157\right)\overset{(50-50\cos\phi)}{\underset{= 0}{-M\phi}}$$

$$-50\int_0^\phi 100\,d\theta\,(\cos\theta - \cos\phi) = 2850\sin\phi - 7850 + 7850\cos\phi + M_\phi$$

$$-5000\int_0^\phi (\cos\theta - \cos\phi)\,d\theta = 2850\sin\phi - 7850 + 7850\cos\phi + M_\phi$$

but

$$\int_0^\phi (\cos\theta - \cos\phi)\,d\theta = \big[\sin\theta - \theta\cos\phi\big]_0^\phi = \sin\phi - \phi\cos\phi$$
$$-0 + 0$$
$$= \sin\phi - \phi\cos\phi$$

$$\therefore \quad -5000\big(\sin\phi - \phi\cos\phi\big) = 2850\sin\phi - 7850 + 7850\cos\phi + M_\phi$$

$$M_\phi = 7850 - 7850\sin\phi - 7850\cos\phi + 5000\phi\cos\phi$$

$$\boxed{M_\phi = 7850 - 7850\sin\phi - 5000\cos\phi\,(1.57 - \phi)} \quad , \quad 0 \le \phi \le \pi$$

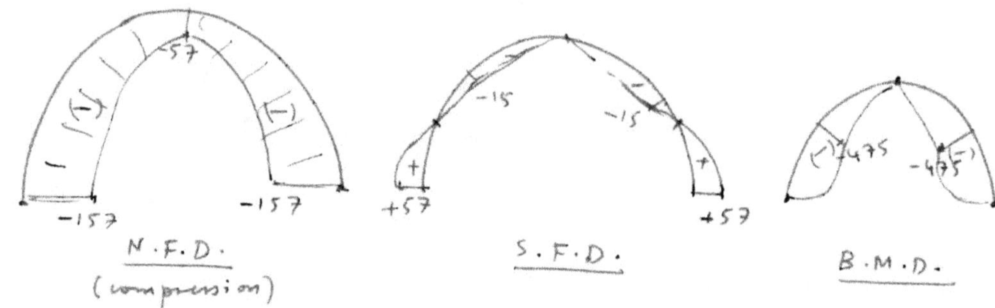

Deflections :

Example :

Calculate the maximum deflection of point A of the curved beam as well as the rotation of the point.

Solution :

$+\circlearrowleft \Sigma M_x = 0:$

$-M_\varphi + P\cdot R\sin\varphi = 0$

$M_\varphi = PR\sin\varphi,$

$\quad 0 \le \varphi \le \pi$.

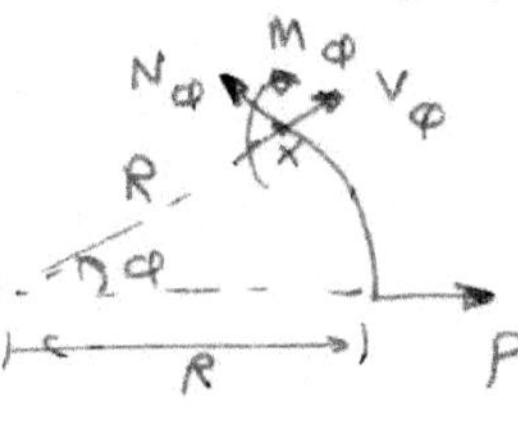

① $m_\varphi = 1\cdot R\sin\varphi = R\sin\varphi$

$$\Delta_{A_H} = \int_S \frac{Mm}{EI}\,ds \quad , \quad \text{but } ds = R\,d\varphi$$

$$= \int_0^\pi \frac{M_\varphi m_\varphi\, R\,d\varphi}{EI} = \frac{1}{EI}\int_0^\pi PR\sin\varphi\cdot R\sin\varphi\cdot R\,d\varphi$$

$$= \frac{PR^3}{EI}\int_0^\pi \sin^2\varphi\,d\varphi = \frac{PR^3}{EI}\int_0^\pi \frac{1-\cos2\varphi}{2}\,d\varphi$$

$$= \frac{PR^3}{2EI}\left[\varphi - \tfrac{1}{2}\sin2\varphi\right]_0^\pi$$

$$= \frac{PR^3}{2EI}\left[\pi - 0\right]$$

$$= \frac{\pi PR^3}{2EI} \; .$$

$\cos2\varphi = \cos^2\varphi - \sin^2\varphi$

$\quad = 1 - \sin^2\varphi - \sin^2\varphi$

$\quad = 1 - 2\sin^2\varphi$

$\sin^2\varphi = \dfrac{1-\cos2\varphi}{2}$

② $+\circlearrowleft \Sigma M_x = 0:$

$-m_\varphi - 1\cdot(R - R\cos\varphi) = 0$

$m_\varphi = -R(1-\cos\varphi) .$

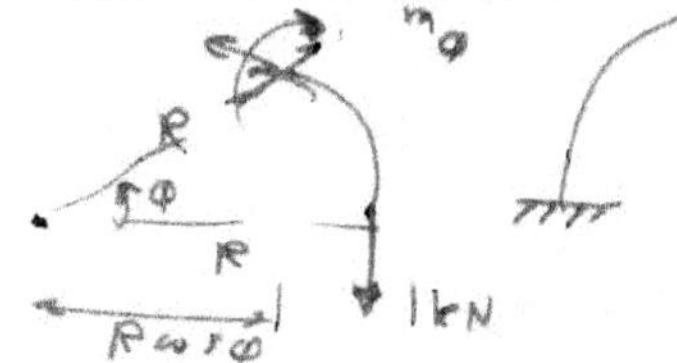

$$\therefore \Delta_{A_v} = \int_0^{\pi} \frac{Mm}{EI} R\, d\phi \;=\; \frac{R}{EI} \int_0^{\pi} PR\sin\phi\,(-R)(1-\cos\phi)\, d\phi$$

$$= -\frac{PR^3}{EI} \int_0^{\pi} (\sin\phi - \sin\phi\cos\phi)\, d\phi \qquad \Big\| \; \sin 2\theta = 2\sin\phi\cos\phi$$

$$= -\frac{PR^3}{EI} \int_0^{\pi} \left(\sin\phi - \frac{\sin 2\phi}{2}\right) d\theta$$

$$= -\frac{PR^3}{EI} \left[-\cos\phi + \tfrac{1}{4}\cos 2\phi \right]_0^{\pi}$$

$$= -\frac{PR^3}{EI} \left[1 - 0 + 1 - 0 \right]$$

$$= -\frac{PR^3}{EI} (2)$$

$$= -\frac{2PR^3}{EI}$$

$$\therefore \Delta_A = \sqrt{\Delta_{A_H}^2 + \Delta_{A_V}^2} \;=\; \sqrt{\left(\frac{\pi PR^3}{2EI}\right)^2 + \left(-\frac{2PR^3}{EI}\right)^2}$$

$$= \frac{PR^3}{EI} \sqrt{\frac{\pi^2}{4} + 4}$$

③ $\quad m_\phi = 1.$

$$\therefore \theta_A = \int_0^{\pi} \frac{Mm}{EI} R\, d\phi$$

$$= \int_0^{\pi} \frac{PR\sin\phi \cdot 1}{EI} R\, d\phi$$

$$= \frac{PR^2}{EI} \int_0^{\pi} \sin\phi\, d\phi$$

$$= \frac{PR^2}{EI} \left[-\cos\phi \right]_0^{\pi} \;=\; \frac{PR^2}{EI}\left[1 + 1 \right] \;=\; \frac{2PR^2}{EI}.$$

<u>Example :</u>

Consider the cantilever curved beam of constant cross-sections and radius R shown in the figure. Assuming that the thickness-to-radius ratio of the beam is very small, calculate the horizontal and vertical deflections at the free end (point B).

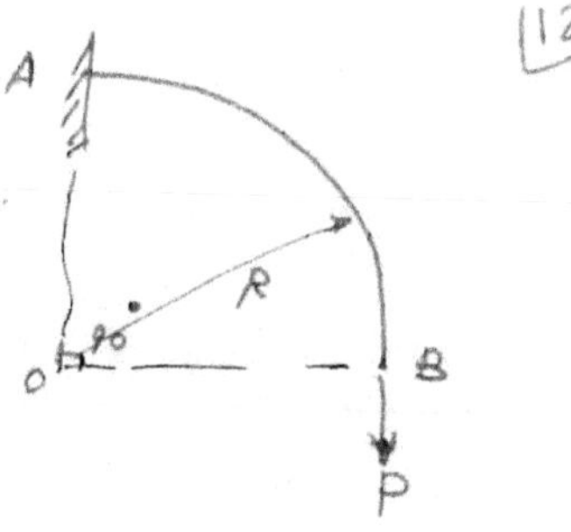

<u>Solution :</u>

$+\!\!\!\int$ $\Sigma M_c = 0$:

$$-M_\phi - P(R - R\cos\phi) = 0$$

$$M_\phi = -PR(1 - \cos\phi) \quad , \quad 0 \leq \phi \leq \frac{\pi}{2}$$

$$m_\phi = -1 \cdot R(1 - \cos\phi) \,.$$

$$\delta_{B_v} = \int_0^{\pi/2} \frac{M_\phi \, m_\phi}{EI} \, ds \quad , \quad ds = R\,d\phi$$

$$= \frac{1}{EI} \int_0^{\pi/2} -PR(1-\cos\phi)\cdot(-R)(1-\cos\phi)\cdot R\,d\phi$$

$$= \frac{PR^3}{EI} \int_0^{\pi/2} (1 - \cos\theta)^2 \, d\theta$$

$$= \frac{PR^3}{EI} \int_0^{\pi/2} (1 - 2\cos\theta + \cos^2\theta)\,d\theta = \frac{PR^3}{EI} \int_0^{\pi/2} \left(1 - 2\cos\theta + \frac{1+\cos 2\theta}{2}\right)d\theta$$

$$= \frac{PR^3}{EI} \int_0^{\pi/2} \left(\frac{3}{2} - 2\cos\theta + \frac{1}{2}\cos 2\theta\right)d\theta$$

$$= \frac{PR^3}{EI} \left[\frac{3}{2}\theta - 2\sin\theta + \frac{1}{4}\sin 2\theta\right]_0^{\pi/2}$$

$$= \frac{PR^3}{EI} \left[\frac{3\pi}{4} - 2 + 0 - 0 - 0 - 0\right]$$

$$= \frac{PR^3}{EI}\left(\frac{3\pi}{4} - 2\right)$$

$$\cos 2\theta = \cos^2\theta - \sin^2\theta$$
$$= 2\cos^2\theta - 1$$
$$\cos^2\theta = \frac{1 + \cos 2\theta}{2}$$

$\circlearrowleft +\ \Sigma M_c = 0:$

$-\cancel{m_\phi} + 1\cdot(R\sin\phi) = 0$

$m_\phi = R\sin\phi$

$\therefore\ \delta_{B_h} = \int_0^{\pi/2} \frac{M_\phi\, m_\phi}{EI}\, ds$

$= \int_0^{\pi/2} \frac{-PR(1-\cos\phi)\cdot R\sin\phi\cdot R\,d\phi}{EI}$

$\sin 2\phi = 2\sin\phi\cos\phi$

$\sin\phi\cos\phi = \tfrac{1}{2}\sin 2\phi$

$= -\frac{PR^3}{EI}\int_0^{\pi/2}(\sin\phi - \sin\phi\cos\phi)\,d\phi$

$= -\frac{PR^3}{EI}\int_0^{\pi/2}\left(\sin\phi - \tfrac{1}{2}\sin 2\phi\right)d\phi$

$= -\frac{PR^3}{EI}\left[-\cos\phi + \tfrac{1}{4}\cos 2\phi\right]_0^{\pi/2}$

$= -\frac{PR^3}{EI}\left[-0 + \tfrac{1}{4}(-1) + 1 - \tfrac{1}{4}\right]$

$= -\frac{PR^3}{EI}\left(1 - \tfrac{1}{2}\right)$

$= -\frac{PR^3}{2EI}\cdot$ (the free end moves to the left).

Influence Lines for Arches:

Example: Consider the three-hinged arch shown in the figure. Draw the influence lines for the reactions at support A and the internal forces (normal force, shear, and moment) at ~~section~~ section 1-1.

Solution:

Reactions:

$$V_A = V_A^0 = \frac{1 \cdot (L-x)}{L}$$

$$V_B = V_B^0 = \frac{x}{L}$$

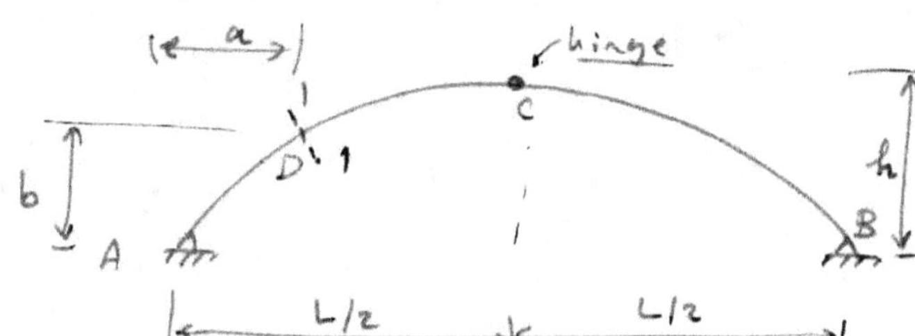

Segments AC & CB:

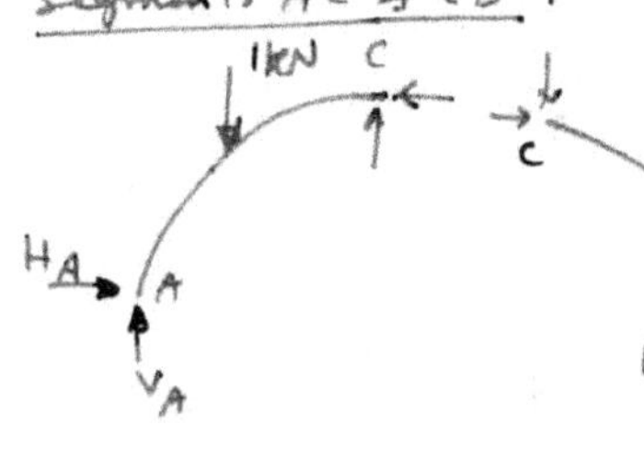

(1): $0 \le x \le L/2$:

$+\circlearrowleft \Sigma M_c = 0$:

$$V_B \left(\frac{L}{2}\right) - H_B \cdot h = 0$$

$$H_B = \frac{L \, V_B}{2h}$$

$$= \frac{L}{2h}\left(\frac{x}{L}\right) = \frac{x}{2h} .$$

(2): $0 \le x \le L$:

$+\circlearrowleft \Sigma M_c = 0$:

$$H_A \cdot h - V_A \cdot \frac{L}{2} = 0$$

$$H_A = \frac{L \, V_A}{2h} = \frac{L}{2h}\left(\frac{L-x}{L}\right)$$

$$= \frac{L-x}{2h} .$$

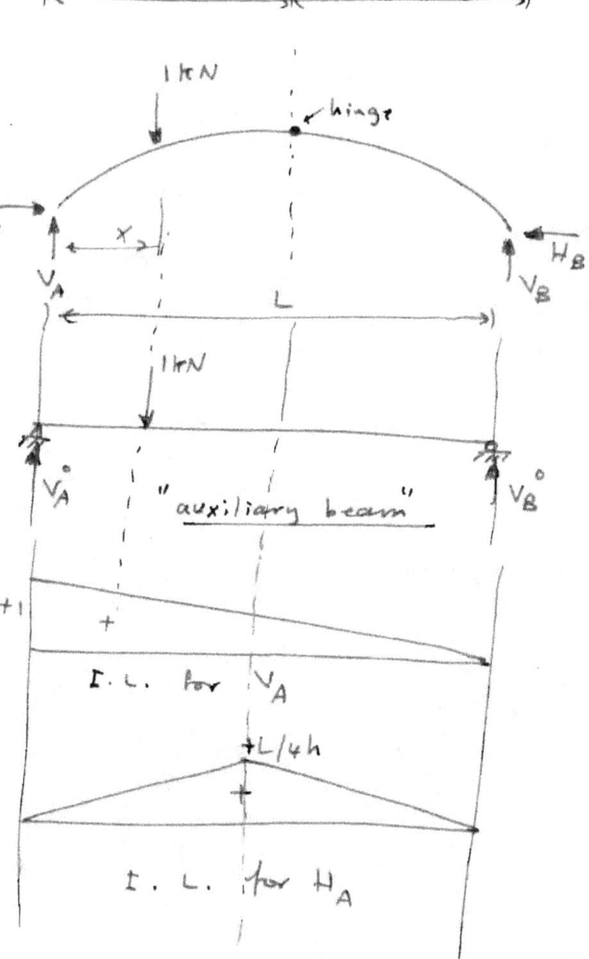

Internal forces :

$\circlearrowleft)\ \Sigma M_D = 0$:

$$M_D + H_A \cdot b - V_A \cdot a + 1 \cdot (a - x) = 0$$

$$M_D = \underbrace{a V_A - 1 \cdot (a - x)}_{M_D^{\circ}} - b H_A$$

$$\boxed{M_D = M_D^{\circ} - b H_A}$$

$\rightarrow \Sigma F_x = 0$:

$$N_D \cos\alpha + V_D \sin\alpha + H_A = 0 \quad\text{——(1)}$$

$\uparrow \Sigma F_y = 0$:

$$N_D \sin\alpha - V_D \cos\alpha + V_A - 1 = 0 \quad\text{——(2)}$$

Solve eqs. (1) & (2) simultaneously for N_D and V_D :

$$N_D \cos^2\alpha + V_D \sin\alpha\cos\alpha + H_A \cos\alpha = 0$$
$$N_D \sin^2\alpha - V_D \sin\alpha\cos\alpha + V_A \sin\alpha - \sin\alpha = 0$$

$$\overline{N_D + H_A \cos\alpha + V_A \sin\alpha - \sin\alpha = 0}$$

$$N_D = (1 - V_A)\sin\alpha - H_A \cos\alpha$$

$$\boxed{N_D = -V_D^{\circ} \sin\alpha - H_A \cos\alpha}$$

Similarly,

$$N_D \sin\alpha\cos\alpha + V_D \sin^2\alpha + H_A \sin\alpha = 0$$
$$-N_D \sin\alpha\cos\alpha + V_D \cos^2\alpha - V_A \cos\alpha + \cos\alpha = 0$$

$$\overline{V_D + H_A \sin\alpha - V_A \cos\alpha + \cos\alpha = 0}$$

$$V_D = (V_A - 1)\cos\alpha - H_A \sin\alpha$$

$$\boxed{V_D = V_D^{\circ} \cos\alpha - H_A \sin\alpha}$$

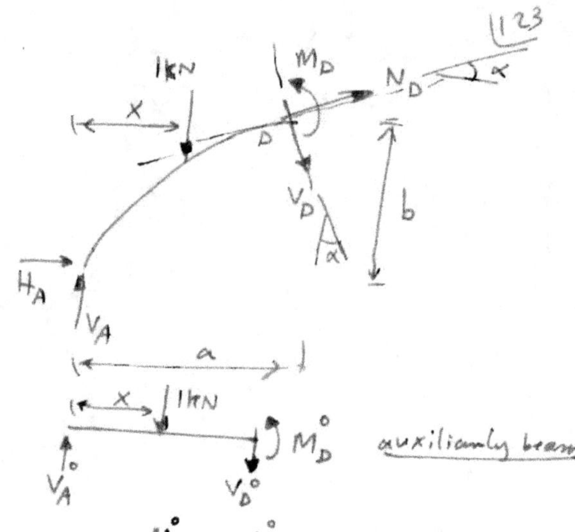

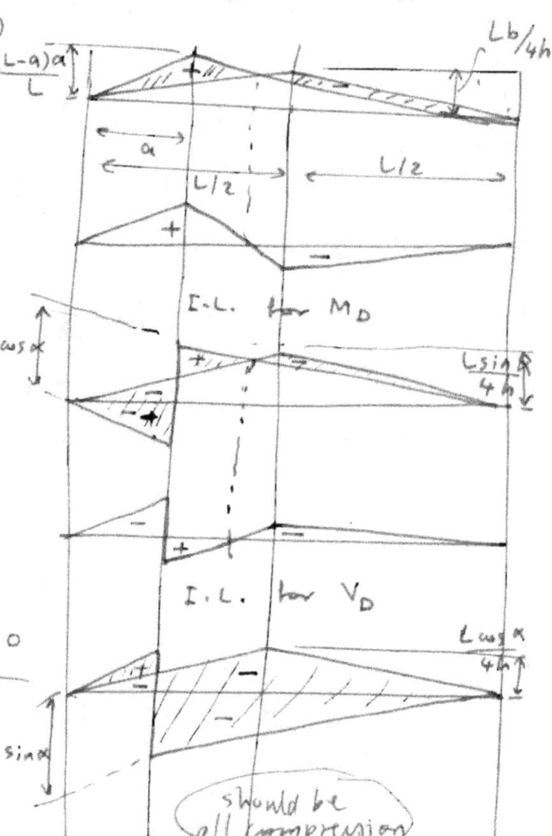

9.2 Cables : (not required)

* The cable is a structural element that resists forces by developing tension stresses.

Example 9.3 :

Determine the reactions and the tension in each segment for the cable shown in the figure.

Solution :

$\circlearrowright \Sigma M_E = 0 :$

$-V_A (60) + 12 (50)$
$+ 12 (25) + 8 (15) = 0$

$V_A = 17 \, kN \uparrow$.

$+\uparrow \Sigma F_y = 0 :$

$V_A + 17 - 12 - 12 - 8 = 0 \implies V_A = 15 \, kN \uparrow$.

Consider part ABC :

$\circlearrowright \Sigma M_C = 0 :$

$H_A (12) - 17 (35) + 12 (25) = 0$

$H_A = 24.6 \, kN \leftarrow$.

For the entire cable :

$\xrightarrow{+} \Sigma F_x = 0 : \quad H_E = 24.6 \, kN \rightarrow$.

Consider segment AB .

Consider joint B :

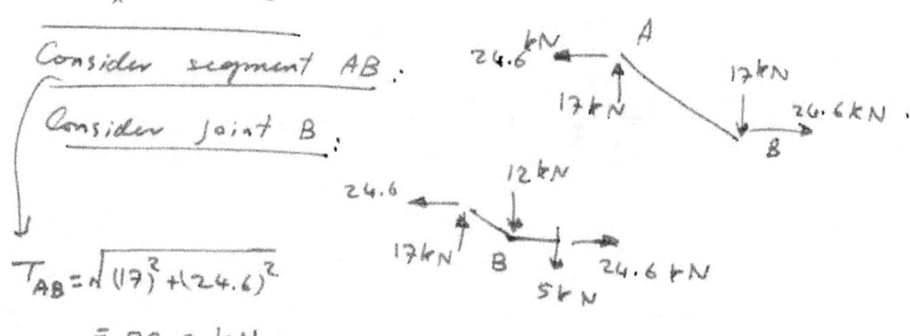

$T_{AB} = \sqrt{(17)^2 + (24.6)^2}$

$\quad\quad = 29.9 \, kN$.

<u>Consider segment BC:</u>

$$T_{BC} = \sqrt{(5)^2 + (24.6)^2}$$
$$= 25.1 \text{ kN}.$$

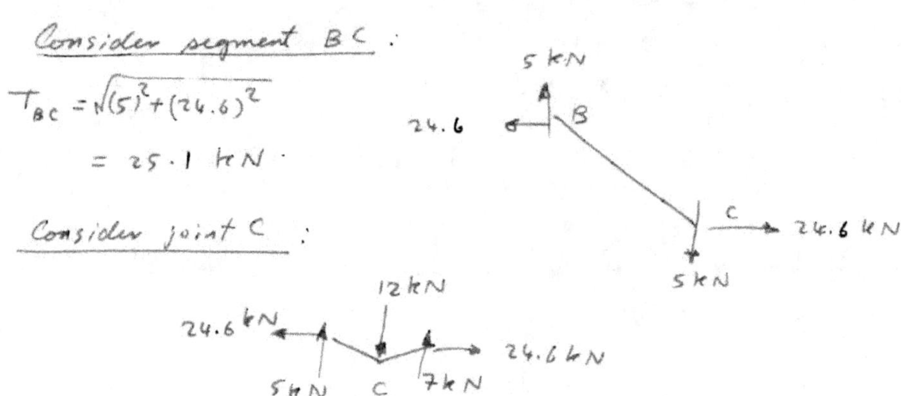

<u>Consider joint C:</u>

<u>Consider segment CD:</u>

$$T_{CD} = \sqrt{(7)^2 + (24.6)^2}$$
$$= 25.6 \text{ kN}.$$

<u>Consider joint D:</u>

<u>Consider segment DE:</u>

$$T_{DE} = \sqrt{(15)^2 + (24.6)^2}$$
$$= 28.8 \text{ kN}.$$

Indeterminate Structures: Introduction

* A structure is <u>determinate</u> if the equations of equilibrium suffice to calculate all the external reactions as well as the internal forces.

* A structure is <u>indeterminate</u> if the equations of equilibrium are insufficient for determining the reactions and internal forces.

Example (1):

Consider the <u>hinged two-bar</u> structure shown in the figure.

$$r = 4$$
$$m = 2 \quad \Big\} \quad m + r = 6 \quad \Big\} \quad \text{stable \& determinate}$$
$$j = 3, \quad 2j = 6$$

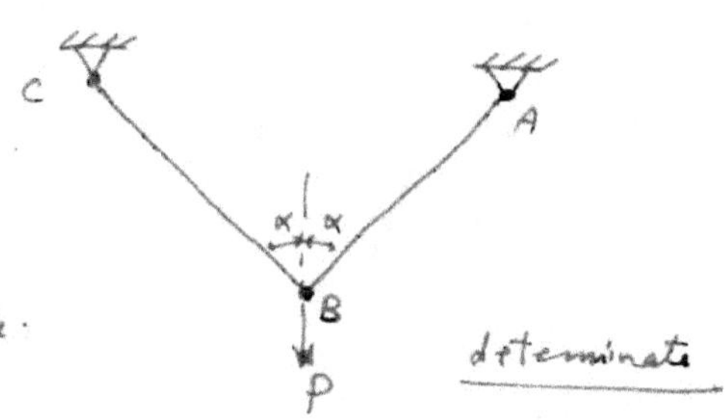

Joint B:

$\xrightarrow{+} \; \Sigma F_x = 0:$

$$F_{AB} \sin\alpha - F_{BC} \sin\alpha = 0$$
$$F_{AB} = F_{BC}$$

$+\uparrow \; \Sigma F_y = 0:$

$$F_{AB} \cos\alpha + F_{BC} \cos\alpha - P = 0$$

but $F_{AB} = F_{BC}$

$$2 F_{AB} \cos\alpha = P \quad \Longrightarrow \quad \boxed{F_{AB} = \frac{P}{2\cos\alpha} = F_{BC}}$$

Example (2):

Consider the <u>hinged three-bar</u> structure shown in the figure.

$$r = 6$$
$$m = 3 \quad \Big\} \quad m + r = 9$$
$$j = 4, \quad 2j = 8, \quad m + r > 2j.$$
$$\therefore \text{ stable \& indeterminate.}$$

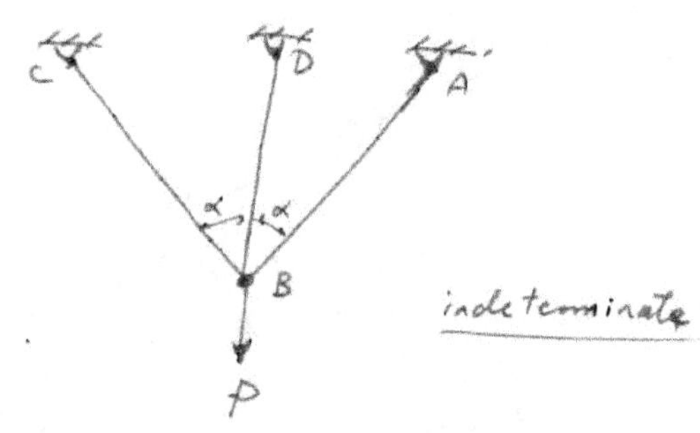

Joint B:

$\rightarrow \Sigma F_x = 0:$

$\qquad F_{AB} \sin\alpha - F_{BC} \sin\alpha = 0$

$$\boxed{F_{AB} = F_{BC}} \qquad\qquad (1)$$

$\uparrow \Sigma F_y = 0:$

$$\boxed{F_{AB} \cos\alpha + F_{BC} \cos\alpha + F_{BD} = P} \qquad (2)$$

Substitute (1) in (2):

$$\boxed{2 F_{BC} \cos\alpha + F_{BD} = P} \qquad\qquad (a)$$

* The available equations of equilibrium are <u>not</u> sufficient to obtain the force in each bar.

* We must consider the deformation (deflections) of the structure in addition to the equations of equilibrium.

$$\boxed{\delta_{BC} = \delta_{BD} \cos\alpha}$$

assuming α is a small angle. ①

but $\quad \delta = \dfrac{PL}{EA}$ (Hooke's law) ②

$$\delta_{BC} = \frac{F_{BC}\, L_{BC}}{EA}$$

$EA \equiv$ constant for all members.

$$\delta_{BC} = \frac{F_{BC}\, \ell}{EA \cos\alpha}$$

$$\delta_{BD} = \frac{F_{BD}\, \ell}{EA}$$

$\therefore \quad \dfrac{F_{BC}\,\ell}{EA\cos\alpha} = \dfrac{F_{BD}\,\ell}{EA} \cos\alpha \implies \boxed{F_{BC} = F_{BD} \cos^2\alpha} \qquad (b)$

Substitute in eq. (a):

$$2\left(F_{BD}\cos^3\alpha\right)\cos\alpha + F_{BD} = P$$

$$(2\cos^3\alpha + 1)F_{BD} = P \quad\Longrightarrow\quad F_{BD} = \frac{P}{1 + 2\cos^3\alpha}$$

$$F_{BC} = \frac{P\cos^2\alpha}{1 + 2\cos^3\alpha} = F_{AB}$$

Types of Arches :

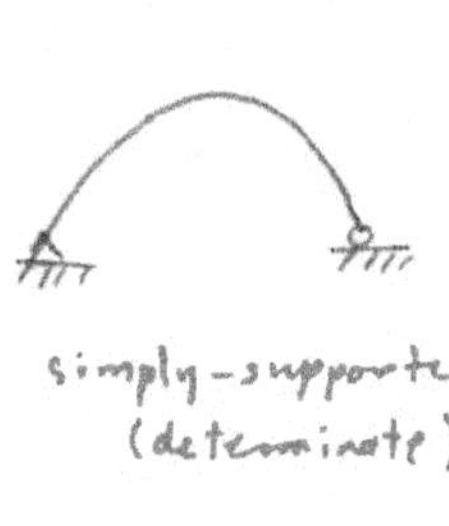
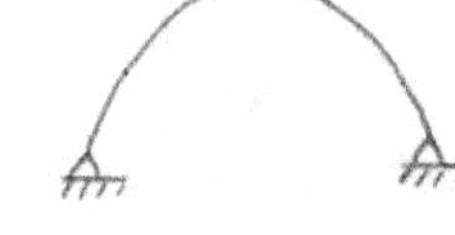
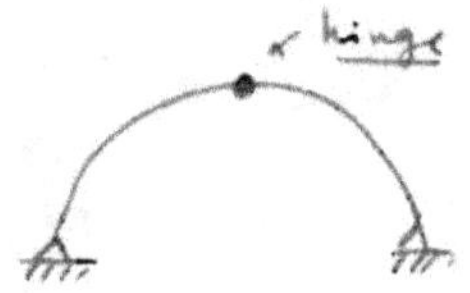

simply-supported arch
(determinate)

two-hinged arch
(indeterminate)

three-hinged arch
(determinate)

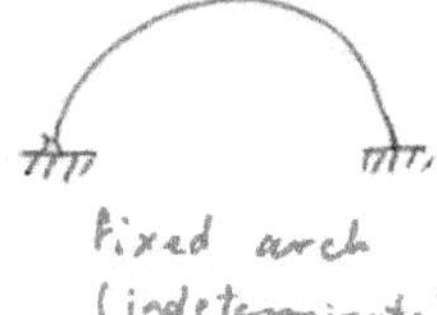
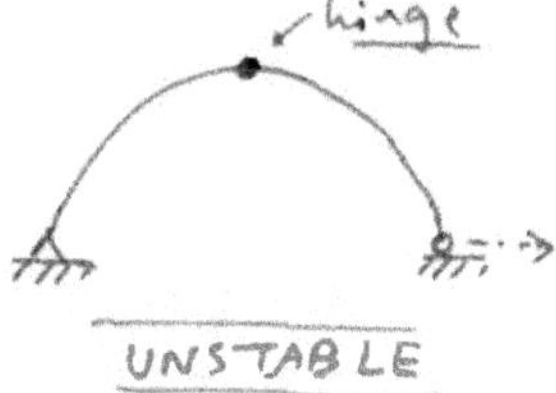
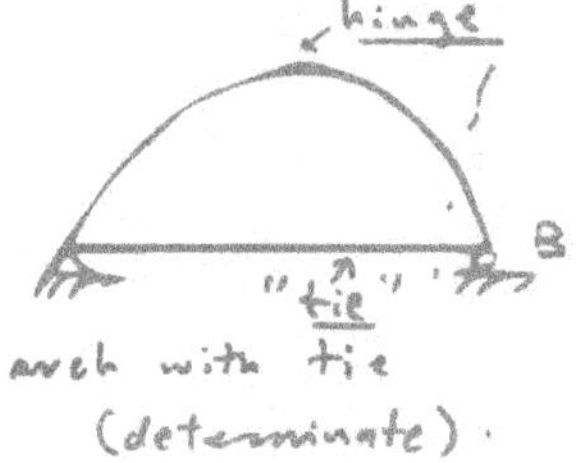

fixed arch
(indeterminate)

UNSTABLE

arch with tie
(determinate).

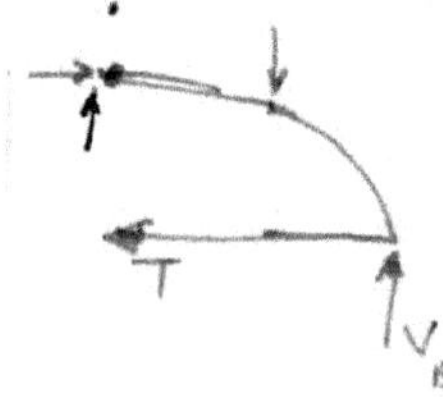

T (tension in the tie) is equivalent to the horizontal reaction H in the ordinary three-hinged arch.

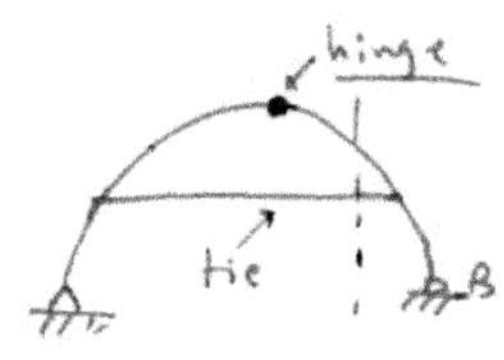
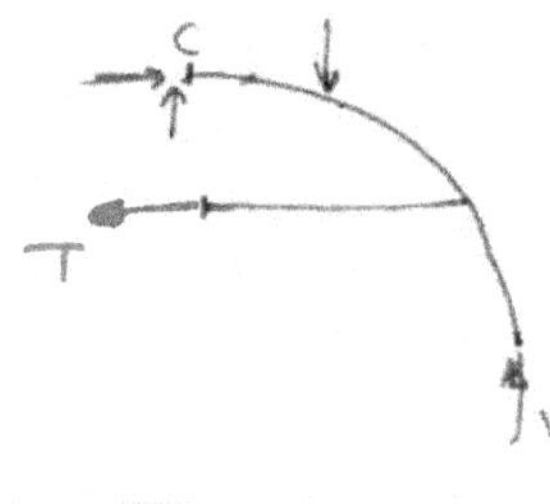

Take moment at the crown to determine T.